Thomas Cairney

BSc., PhD., CEng., MICE, MIWES

Hydraulics for civil engineering technicians

Longman London and New York

Longman Group Limited
Longman House, Burnt Mill, Harlow
Essex CM20 2JE, England
Associated companies throughout the world

Published in the United States of America
by Longman Inc., New York

© Longman Group Limited 1984

First published 1984

British Library Cataloguing in Publication Data
Cairney, T.
   Hydraulics for civil engineering technicians.
   — (Longman technician series: construction
   and civil engineering sector)
   1. Hydraulic engineering
   I. Title
   627   TC145

ISBN 0-582-41303-6

Set in 10/11 pt Lasercomp Times
Printed in Singapore by Selector Printing Company.

Longman Technician Series

Construction and Civil Engineering

General Editor — Construction and Civil Engineering

C.R. Bassett, B.Sc., F.C.I.O.B.

Formerly Principal Lecturer in the Department of Building and Surveying, Guildford County College of Technology

Books published in this sector of the series:

Building organisations and procedures *G. Forster*
Construction site studies – production, administration
    and personnel *G. Forster*
Practical construction science *B. J. Smith*
Construction surveying *G. A. Scott*
Materials and structures *R. Whitlow*
Construction technology Volume 1 *R. Chudley*
Construction technology Volume 2 *R. Chudley*
Construction technology Volume 3 *R. Chudley*
Construction technology Volume 4 *R. Chudley*
Maintenance and adaptation of buildings *R. Chudley*
Building services and equipment Volume 1 *F. Hall*
Building services and equipment Volume 2 *F. Hall*
Building services and equipment Volume 3 *F. Hall*
Measurement Level 2 *M. Gardner*
Measurement Level 3 *M. Gardner*
Structural analysis Level 4 *G. B. Vine*
Site surveying and levelling Level 2 *H. Rawlinson*
Economics for the construction industry *R. C. Shutt*
Design procedures Level 4 *J. M. Zunde*
Design technology Level 5 *J. M. Zunde*
Architectural design procedures *C. M. H. Barritt*
Construction technology for civil engineering technicians
    P. L. Monckton

Chapter 6   Methods of analysing fluid flows   62

Chapter 7   Gravity flow in pipes   84

Chapter 8   More complex gravity pipes   108

Chapter 9   Unsteady flow in pipes 128

Chapter 10   Hydraulic machinery 143

Preface

This book has been written for Civil Engineering students, who have just commenced Degree or Technician Education Council courses.

All too often such people find Hydraulics one of the more difficult areas of study.

The basic reason for this is not that the subject is particularly hard to understand, but that its various sub-divisions can be treated to such degrees of complexity that they assume the status of subjects in their own right. This approach frequently confuses students, who are unable to correlate the different sub-divisions and so utilise hydraulic principles in civil engineering design and analysis – surely the main reason for a Civil Engineer to study any subject.

In an effort to remedy this situation, the present book has been made as readable and as simple as is possible, and uses self-assessment questions and worked examples to ensure that students understand each section of the subject before proceeding.

Obviously some depth of treatment has been lost by this approach, but this is not believed to present any real problem. There are many excellent and detailed text-books available which a student can consult, once he or she has grasped the basic principles and inter-relationships of the subject.

Hydraulic Engineering was a crucial foundation of the birth of civilised society and still offers a rewarding and important career. If this introductory text helps even a proportion of students to appreciate this then the effort of producing the book will have been justified.

My thanks must be given to my departmental colleagues who advised on parts of the book, and are particularly due to Miss J. Adams and Miss K. Lovatt who uncomplainingly carried out the typing and drafting.

T. Cairney
Department of Building and Civil Engineering
Liverpool Polytechnic

Acknowledgements

We are indebted to the following for permission to use copyright material:

Hydraulics Research Station Limited, Wallingford, Oxfordshire for our Fig. 7.7; Mr N. Traynor, Liverpool Polytechnic for our Table 8.1.

Chapter 1

The value of hydraulics

1.1 What is hydraulics?

Fire has been said to be a good servant, but a bad master. The same can be said of water. Floods and droughts are periods of human misery, marked by losses of life, property and opportunity. Those countries which have yet to control their waters are typified by endemic diseases, a lack of development and a limited life expectancy.

The basic aim of hydraulics is thus to understand, and so control for the benefit of society, the occurrence, movement and use of water, whether it is in lakes, rivers, pipes, drains, percolating through soils or pounding the coastline as destructive waves. To modify the behaviour of water calls inevitably for a large investment of time, resources and effort. Thus hydraulic engineering has only appeared once a society is centralised under an organised government. This point will be obvious if one considers how much effort and time it would take one man to dig even a narrow canal and how impossible this would be if he had to clothe, feed and shelter himself at the same time.

Hydraulics is often confused with the allied science of fluid mechanics. Good reason for this exists, since a considerable overlap occurs between the two studies. However, fluid mechanics deals with gases, as well as the common liquids, and to most civil engineers a study of gas behaviour is irrelevant to their professional needs.

Thus the modern definition of hydraulics limits the subject to the study of water and those other liquids which a civil engineer is called upon to store, convey or pump.

1.2 Why do we study hydraulics?

All organised societies need adequate water supplies, drainage to dispose of waste or excess water, as well as the protection from uncontrolled water. Thus an obvious necessity for a study of hydraulics exists.

However, this self-evident need does tend to persuade many students that hydraulics is separate from, and often more difficult than, the other major civil engineering studies – structural mechanics, structural design and soil mechanics. This is an unfortunate belief, and can lead to the attitude that hydraulics is irrelevant to (for example) structural projects. Yet there are outstandingly successful Roman projects (some of which still function after 1 700 years) which have survived because their designers knew that, if water could penetrate into the structures, failure would be inevitable in a few years.

Some of the examples in the following chapters will stress the interaction of hydraulic factors with other civil engineering studies. Most, however, will be limited to illustrating the behaviour of fluids, simply because a relatively short book has to cover its main subject matter. Thus it is important that the reader realises that hydraulic problems do not occur in isolation and that in professional life it will be essential to integrate hydraulics with the other academic subjects.

1.3 The development of hydraulics

Hydraulic engineering has been practised for as long ago as written history.

The ancient Egyptians, Babylonians and Chinese constructed canals, dams and devices for lifting water. Whilst some of these works were very successful, others are known to have failed. The lack of any comprehensive theory of hydraulics made any of their major hydraulic projects something of a gamble.

By the time Greek civilisation had become established (between 500 and 100 BC) enough information had been collected to divide Hydraulics into Hydrostatics (the study of motionless fluids) and Hydrodynamics (the science of moving fluids). In hydrostatics, the only force acting is the weight of the body of liquid, and with such a simple situation the Greeks were able to establish almost every rule and application. The effect of this was the production of a variety of machines operated by water.

Hydrodynamics, however, is complicated by other forces, and even today it is recognised as a more complex study. The Greeks, biased as they were against experimentation, were unable to understand its complexity.

The Roman Empire followed the Greek civilisation and was obviously marked by a vast upsurge in hydraulic engineering. The empire covered much of Europe, Asia and northern Africa, and was studded with public water supply schemes, drainage works and bridges, many of which still stand today. There is no doubt that the Romans were competent appliers of hydraulics but, despite this, there is evidence that they lacked any deep understanding of the science. A good example of this is their method of charging for private water supplies to the wealthier citizens. Their approach was to believe that the flow rate delivered depended only on the diameter of the pipe used for the supply and thus the water charges were based on the pipe size. Today we appreciate that a flow rate is expressed in units such as cubic metres per second (m^3/s) and that it is affected not only by the area open to flow (square metres: m^2) but also by the velocity of flow, in metres per second (m/s).

Without a real understanding of what flow actually was, it is not surprising that the Romans contributed little to our understanding of hydrodynamics. The amazing thing is that their works proved as successful as they did.

Until the sixteenth century, little real progress occurred. Then with the Renaissance, spreading out from Italy, for the first time it became common for intelligent men to improve their understanding by conducting experiments with real flows. This growth of experimental knowledge, combined some 200 years later with a renewed interest in mathematical analysis by such workers as Newton, Pascal and Descartes, led to the start of a well established hydrodynamic theory.

Since then, hydraulics has developed and has become a more exact science. Perhaps the major cause of this was the Industrial Revolution with its vast demand for water supplies, drainage and water-powered machines. The businessmen who controlled the new industrialisation demanded that civil engineers should supply exactly the water or drainage or power that was required, and so forced the development of more precise design methods.

Today hydraulic engineering has reached the stage of confidence that it is possible to rechannel major rivers, to develop hydroelectric power adequate to supply a small country's needs and to build ports and breakwaters on coasts where it was formerly impossible to dock more than a small canoe.

1.4 The difficulties a student encounters

Hydraulics is still divided into the two categories the Greeks recognised – hydrostatics and hydrodynamics.

Hydrostatics is always the first part studied and usually occupies less than one third of the total time. As it deals only with a single type of force – due to the weight of the fluid in the tank or behind the dam wall – it resembles closely the study of solid body mechanics and utilises much the same methods. For example, it is often necessary in the hydrostatic design of (say) a lock gate to take the moments of the various fluid forces about a point, just as one would do for loads in structural mechanics. Thus any student having difficulties with hydrostatic problems would be well advised to revise the basics of applied mechanics.

Hydrodynamics is the largest and certainly the most interesting part of hydraulics. It does, however, create real problems for some students, who do not recognise that the laboratory experiments are as necessary to an overall understanding as is the published theory. Hydrodynamics is a complex subject, and everyday life offers very little opportunity to become familiar with some of its important effects. For example, the boundary layers, total energy and hydraulic grade lines in pipe flow, the variable depths of flow that can occur in open channels and the remarkably high, yet short-lived, pressures that occur when a valve on a pipeline is closed rapidly, are all difficult to understand unless one can visualise them. Laboratory experiments and demonstrations are the only possible chance for this. Thus laboratory periods are at least as important as theoretical classes and should always be taken as the opportunity to understand what is happening in the fluid system.

The other problem that students commonly encounter with hydrodynamics is that of integrating the various sub-topics that are covered. The only reason for studying hydraulics is surely to be able to apply the subject matter to real life problems and that requires above all else an overall grasp of the material. Details of the specific formula for (say) the flow rate over a particular type of weir can always be found in any standard text book. What no book can give you is a personal integration of the subject which will allow you to identify what analytical or design technique is necessary in a particular part of a hydraulic

If these two common mistakes are avoided, and if it is always realised that hydraulics is an integral part of civil engineering, the subject can usually be studied with, at least, a fair degree of enjoyment and certainly with a sense of personal achievement.

The ability to control water for the advantage of the community has historically been highly prized and even today this is still the case.

Self assessment questions

1. Is it true that hydraulic engineering requires a major investment of a society's resources? Find out how long it took to construct any major scheme (for example, the Suez or the Panama Canal).
2. Explain why a modern society needs hydraulic engineering.
3. Consider how far hydraulics is an important factor in other types of civil engineering schemes. Give examples.
4. Identify the two major sub-divisions of hydraulics. What is the essential difference between these two?
5. Look at any suitable book in your college library and decide whether or not the Roman Empire was able to carry out major hydraulic engineering works.
6. Hydraulics always needs a combination of experimental results and theoretical methods. Is this true?

Chapter 2

Fluid pressures

2.1 Pressure intensity, total pressure, absolute pressure

2.1.1 What is fluid pressure?

Everyday experience proves that pressures increase with depth in a fluid. To swim at the bottom of the deeper end of a swimming pool, under perhaps 2 metres of water, is certainly more painful on the swimmer's ears than if he or she were swimming just below the surface of the water.

The question then is what exactly is meant by the phrase 'fluid pressure'?

Perhaps the easiest way to visualise fluid pressures is to consider the forces that act on a body, 1 metre long by 1 metre wide, floating at a depth of h metres below the surface of a fluid. As the body is neither rising nor sinking, it must be in equilibrium with whatever external forces exist.

Of these, the only one that can act on the body's upper surface is the weight of the column of fluid above the body. This weight obviously is the volume (in cubic metres) of the fluid column times the weight of a cubic metre of the particular fluid.

Thus, the force on the upper surface of the body,

= the weight of the fluid column it supports,

= the volume of the fluid column times the unit weight of the fluid,

= the base area of the column, times the height of the column, times the unit weight of the fluid,

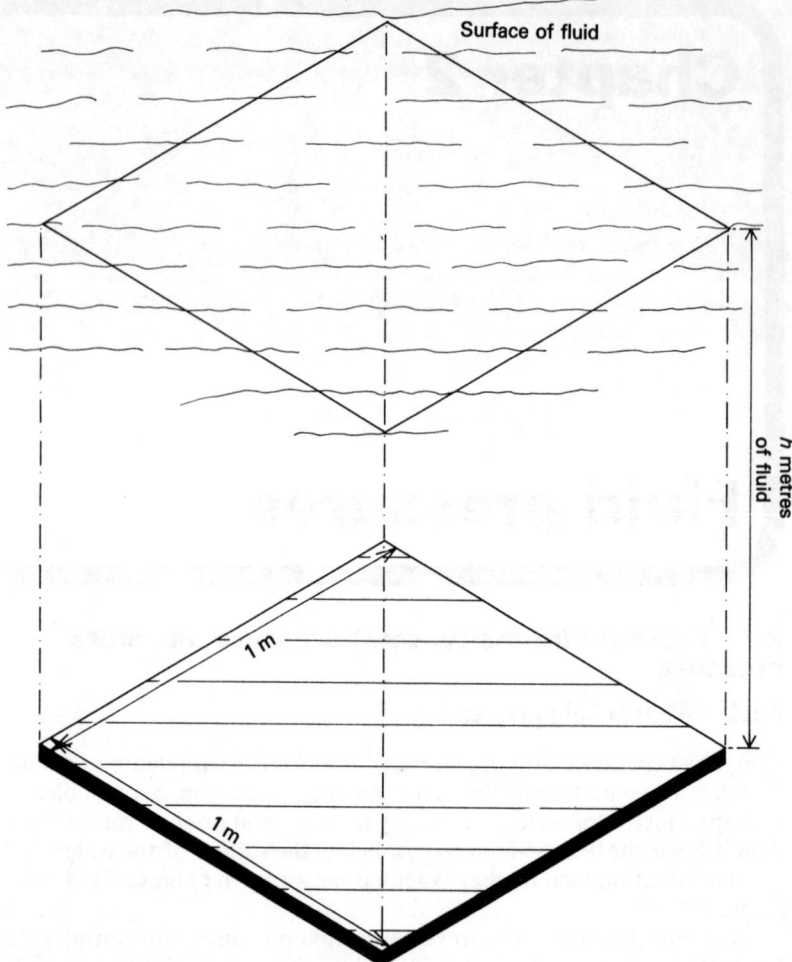

Fig. 2.1 Fluid pressure on a plate

$= (1 \times 1) \times (h) \times$ (the unit mass of the fluid times the gravitational attraction)

$= (1 \times 1) \times (h) \times (M \times g)$.

But, the density of the fluid, by definition, is the mass per unit volume,

i.e. $\rho = \dfrac{M}{V}$

$\therefore \quad M = \rho \times V$

So the force

(or the fluid pressure) $= (1 \times 1) \times (h) \times (\rho \times g)$

$$= \rho \times g \times h$$

Thus *the* fluid pressure on the body is simply the force due to the weight of fluid above it and can be expressed by the equation

$$P = \rho \cdot g \cdot h \qquad [2.1]$$

This equation is the basic tool for all work involving fluid pressure measurement and analysis and is of importance not only in hydraulics but also in geotechnical, structural and foundation engineering.

The units of the fluid force (P) can be found by inserting the units of the terms on the right-hand side of [2.1],

i.e. $P = \rho \times g \times h$

$$= \left(\frac{kg}{m^3}\right)\left(\frac{m}{s^2}\right)(m)$$

$$= \frac{kg}{m \cdot s^2}$$

$$= \left(\frac{kg \cdot m}{s^2}\right)\left(\frac{1}{m^2}\right)$$

$$= \frac{\dfrac{kg \cdot m}{s^2}}{m^2}$$

$$= \frac{newtons}{metre^2} \; (N/m^2)$$

Therefore, the basic equation gives a measurement of force per unit area, or, to restate the result, it expresses a

pressure intensity

2.1.2 Total force

If the total force on a surface is wanted, it is obtained simply by multiplying the calculated pressure intensity by the area of the surface on which the force acts, i.e.

the **total pressure** = the pressure intensity × the area of the surface

$$= \frac{newtons}{metre^2} \times metre^2$$

$$= \mathbf{newtons} \; (N)$$

Whilst pressure intensity and total pressure are commonly expressed as

newtons/metre^2 (N/m^2) and
newtons (N) respectively

other convenient units for larger values are

kilonewtons/metre² (kN/m²)   and
kilonewtons (kN)

where the kilonewton equals 1000 newtons.

In some applications it is also convenient to use square millimetres (mm²) as the unit of area for pressure intensities. One million millimetres² (10^6 mm²) are equivalent to one metre squared (m²).

Although professional engineers tend to use the word 'pressure' rather loosely to express either pressure intensity or total pressure, the context of the argument and the units in which the pressure value is given always distinguish the type of pressure. Students, relatively unfamiliar with the subject, should however always specify precisely which type of pressure is meant to ensure that a confusion in calculation does not occur.

2.1.3   Absolute pressure

So far, the effect of atmospheric pressure has been ignored. If *absolute pressure* is required, this is simply obtained by adding the local value of atmospheric pressure intensity to the *gauge value* (i.e. to the calculated or measured value of pressure intensity). In practice, absolute pressures tend to be relatively unpopular, not only because the atmospheric pressure varies from place to place (it is of course widely appreciated that atmospheric pressure is greatest at sea level and that it declines with height above the sea), but also because it varies at any one place with changes in the local weather. An important point to note is that atmospheric pressures are measured in a variety of units (bars, millibars, millimetres of mercury column, millimetres of water etc.) and it is essential to ensure that units consistent with those of the gauge values are used. The bar (equivalent to 10^5 N/m²) and the millibar (equivalent to 10^2 N/m²) are perhaps the most widely used units for atmospheric pressure intensity. Units in lengths of columns of liquids, such as mercury, can be expressed in newtons/metre² simply by using [2.1] and inserting the appropriate values for the column length and the density of the liquid in the column.

Example 2.1

Convert a barometric pressure of 1010 mm of mercury to its equivalent value in pressure intensity terms.

Solution:

Since $P = \rho \cdot g \cdot h$

$$= (13.6 \times 1000)(9.81)(1.010)$$

since mercury is 13.6 times as dense as water

$$= 134.75 \times 10^3 \text{ N/m}^2$$
$$= 134.75 \text{ kN/m}^2$$

Section 2.1 – Self assessment questions

1. Is a pressure of 132.55 N/m^2 a measure of the pressure intensity or of the total pressure?
2. If the pressure given in Q. 1 acts on an area of 4.500 m^2, what is the total pressure?
3. What fluid pressure intensity would be caused by covering a plate with a 1 metre depth of fresh water (density, 1000 kg/m^3)?
4. Express the answer to Q. 3 in (a) N/mm^2; (b) millibars.
5. Does the pressure intensity on a diver differ if he is immersed under 18 m of (a) fresh water, (b) sea water? Take the density of fresh water as 1000 kg/m^3 and that of sea water as 1019.37 kg/m^3.
6. What depth of water corresponds to a pressure intensity of 340 kN/m^2?
   Assume the water is fresh.
7. A mass (50 kg) lies on a plate (surface area 100 cm^2) and the plate floats on water. What is the total pressure on the plate? What pressure intensity is transmitted to the water below the plate?
8. A plate lies horizontally at a depth of 6.500 m below the surface of a fluid.
   (a) What difference in pressure intensity would occur if the fluid initially was fresh water, which was later replaced by oil?
       ($\rho_{oil} = 750$ kg/m^3).
   (b) In each case what would the total pressure be? Assume the plate's area is 14.350 m^2.
   (c) If the barometric pressure were equivalent to 750 mm of mercury what would the absolute pressure be in each case?
9. At the foot of a mountain, a mercury barometer reads 740 mm and the same barometer taken to the top of the mountain reads 590 mm. If the density of air is constant at a value of 1.225 kg/m^3, what is the height of the mountain in metres? (*Hint:* The difference in mercury barometer readings is due to the different lengths of air column above the base and top of the mountain.)

2.2 Pressure variation with depth

Equation [2.1] confirms the everyday experience that fluid pressure intensity increases with depth of fluid. Further, the equation shows that the increase is linear.

The case of a dam wall holding back 10 metres depth of water (Fig. 2.2) illustrates this. The diagram of pressure intensities is triangular, with individual values varying from zero at the water surface to a maximum of 98.1 kN/m^2 at the base of the dam wall.

For many practical purposes, it is convenient to replace these differing pressure intensities by a total fluid thrust on the wall. This is found quite simply by deciding the area of the wall on which the pressure intensities are deemed to act, and it is usual to consider the total thrust

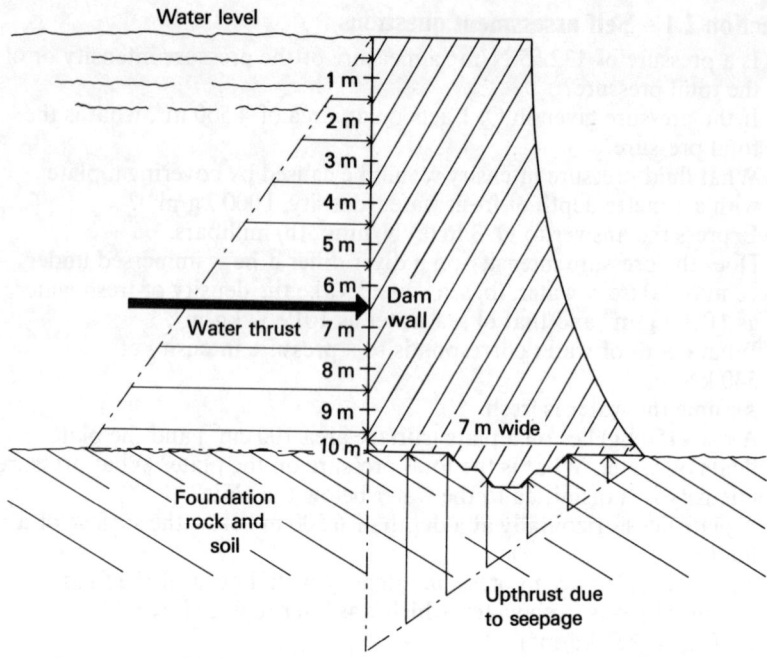

Fig. 2.2 Hydrostatic pressure and uplifts on a dam wall

on each metre length of wall. The area of effect would therefore be
1 metre long by (in this case) 10 metres high, or 10 m². The pressure
intensities, varying as they do from zero to a maximum value of ρgh
at the base of the wall, have to be averaged, and the

$$\begin{pmatrix} \text{total fluid thrust} \\ \text{on the dam wall} \end{pmatrix} = \begin{pmatrix} \text{the average} \\ \text{pressure intensity} \end{pmatrix} \times \begin{pmatrix} \text{the area on which} \\ \text{this average fluid} \\ \text{pressure intensity acts} \end{pmatrix}$$

$$= \frac{\begin{pmatrix} \text{pressure intensity} \\ \text{at water surface} \end{pmatrix} + \begin{pmatrix} \text{pressure intensity} \\ \text{at dam base} \end{pmatrix}}{2} \times \begin{pmatrix} \text{area on which} \\ \text{the average fluid} \\ \text{pressure intensity acts} \end{pmatrix}$$

$$= \frac{0 + \rho gh}{2} \times (h \times 1)$$

$$= \frac{\rho gh^2}{2} \qquad\qquad\qquad [2.2]$$

$$= \text{in this case, } \frac{1000 \times 9.81 \times 10^2}{2}$$

$$= \underline{490.5 \text{ kN on each m length of wall.}}$$

This total fluid thrust can – as will be shown later – be taken as a single force acting at a point one third of the water depth, above the base of the dam.

2.3 Pressure variation with direction

It has already been shown that the fluid pressure intensity at any point simply results from the weight of fluid column above that point. Thus the horizontal surface of a still fluid proves that the atmospheric pressure intensities on the fluid surface must be the same at all points.

Similarly the fluid pressure intensities along any horizontal plane in a still fluid are identical in value. Use will be made of this basic fact when the analysis of pressure-measuring devices is considered.

Even more important, it can be shown that the pressure intensity at any point in a fluid is the same in all the directions around the point. It is worth understanding the proof of this, since a practical consequence important in foundation engineering emerges.

Consider a minute right-angled prism of fluid ABC, within a much larger body of still fluid (Fig. 2.3). The pressure intensities on the three sides are P_a, P_b and P_c. Thus the total pressure forces on the sides are simply the product of the individual pressure intensity and the area of the side it acts upon.

If only a thin slice of the prism is analysed, so thin that its thickness is negligible, then the forces on its triangular cross-section are

$P_a(AB)$, $P_b(AC)$ and $P_c(CB)$,

where the terms in (brackets) indicate the lengths of the sides.

The mass of the thin slice of the prism of fluid is the area of the slice times its density, i.e.

$\frac{1}{2}(AC)(AB) \times \rho$ (kilograms)

and its weight is $\frac{1}{2}(AC)(AB) \times \rho \times g$ (newtons)

Since the prism is neither rising nor sinking, it must be in equilibrium with the surrounding forces. These forces can be resolved into their horizontal and vertical components – *thus* – in the horizontal plane

$P_b(AC) - P_c(CB)(\sin \phi) = 0$

But simple trigonometry shows that

AC = CB sin ϕ

$\therefore \quad P_b = P_c$

in the vertical plane

$P_a(AB) - P_c(BC)(\cos \phi) - \frac{1}{2}g(AC)(AB) = 0$

But, as above, simple trigonometry shows that

AB = BC cos ϕ

$\therefore \quad P_a = P_c + \frac{1}{2}gAC$

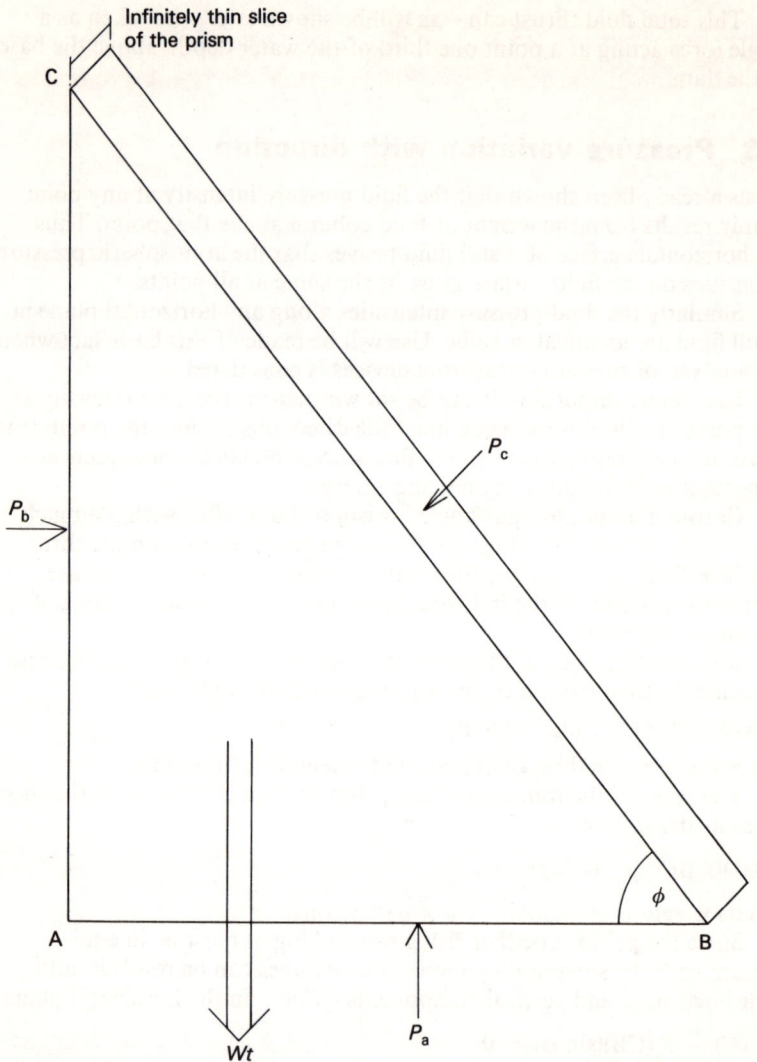

Fig. 2.3 Pressure at a point

and if the prism is very small indeed the length of the side AC tends to a minute value and

$$P_a = P_c$$
$$\therefore \quad P_a = P_b = P_c$$

or

The calculated pressure intensity at any point in a liquid is the same, no matter the direction in which it is deemed to act.

   This result is of particular importance in the design of water-retaining structures such as dams or the walls of tanks. Despite the advances in modern construction techniques it is not possible to build a dam wall which is totally leak proof. Some water always percolates either under the dam's foundations or through minute cracks in the dam wall.

   Thus in Fig. 2.2 the dam holds 10 m of water on one side and none on the other. If leakage does occur below the dam, the water entering under the upstream face has a pressure intensity of 98.1 kN/m^2 (i.e. ρgh, where $h = 10$ m), due to the height of water column above the base. At the downstream face, the emerging leakage is under no hydrostatic pressure and has a gauge intensity of zero.

   The fact that pressure intensities at a point are the same, irrespective of direction, means that the dam wall has a hydrostatic force acting upwards on the base of the structure. This force can be quantified in exactly the same way as that of the fluid thrust on the dam wall (section 2.2) by determining the average uplift pressure intensity and the area on which it acts.

   In this case, this is

$$\begin{pmatrix} \text{the average} \\ \text{pressure intensity} \end{pmatrix} \times \begin{pmatrix} \text{the area of the dam} \\ \text{on which it acts} \end{pmatrix}$$

$$\left(\frac{\rho gh + 0}{2}\right) \times \begin{pmatrix} \text{1 metre wide length of dam} \\ \text{times the width of the dam base} \end{pmatrix}$$

$$= \left(\frac{\rho gh}{2}\right) \times (1 \times w)$$

$$= \frac{1000 \times 9.81 \times 10}{2} \times 7$$

$$= \underline{343.350 \text{ kN, per metre length of dam wall}}$$

   In general, the equation for the uplift force (U) due to seepage under or through a dam wall is

$$U = \rho \cdot g \cdot \left(\frac{h1 + h2}{2}\right)(w) \qquad\qquad [2.3]$$

where w = the dam width in metres.

   Uplift forces also occur where structures have to be founded on granular materials, which lie below the groundwater level, at least, at some times of the year. Such sites were often found unsatisfactory by earlier engineers and consequently are often the only unoccupied land in many modern towns.

   In the wetter part of the year, the excess rainfall seeps into the ground and fills the granular material, often to a considerable height above the level that occurs in the drier months. Fluid pressure intensities

14

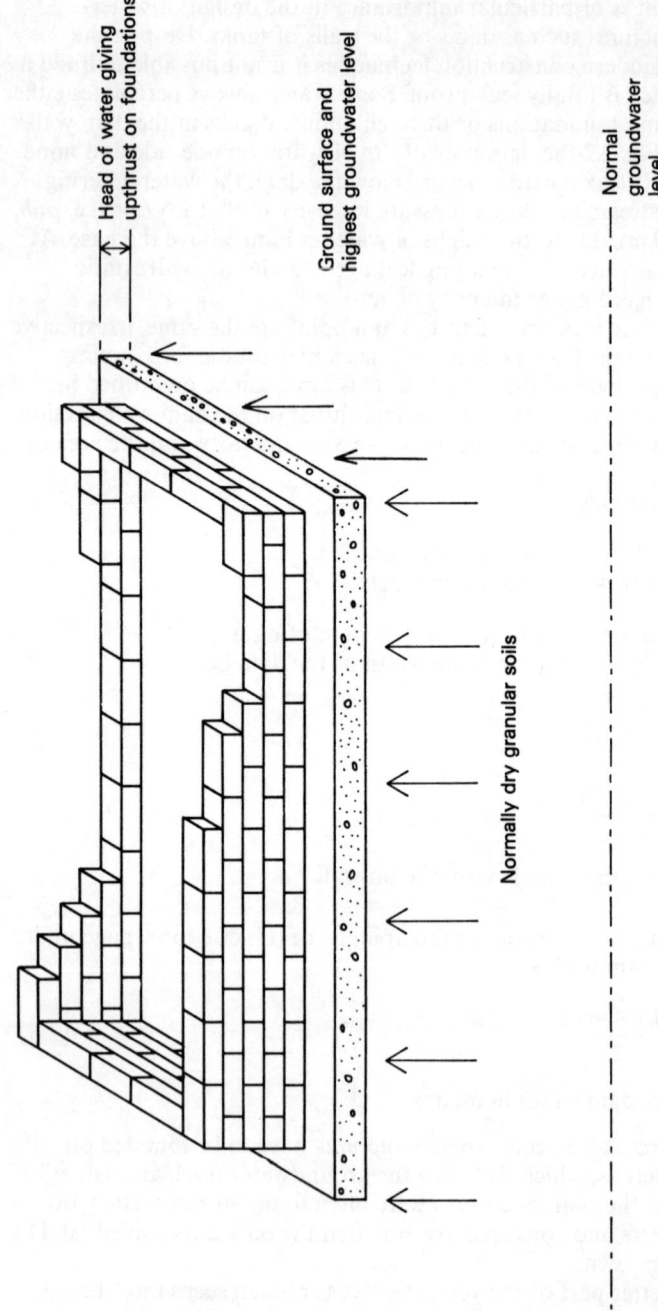

Head of water giving upthrust on foundations

Ground surface and highest groundwater level

Normally dry granular soils

Normal groundwater level

Fig. 2.4

inevitably increase as the depth of groundwater increases and quite significant uplift forces can occur on the foundation (Fig. 2.4).

The pressure intensities are, of course, identical on the base of the foundation slab, as each results from the same depths of groundwater. Thus the total uplift is simply the

(area of the foundation slab) times (the pressure intensity)   [2.4]

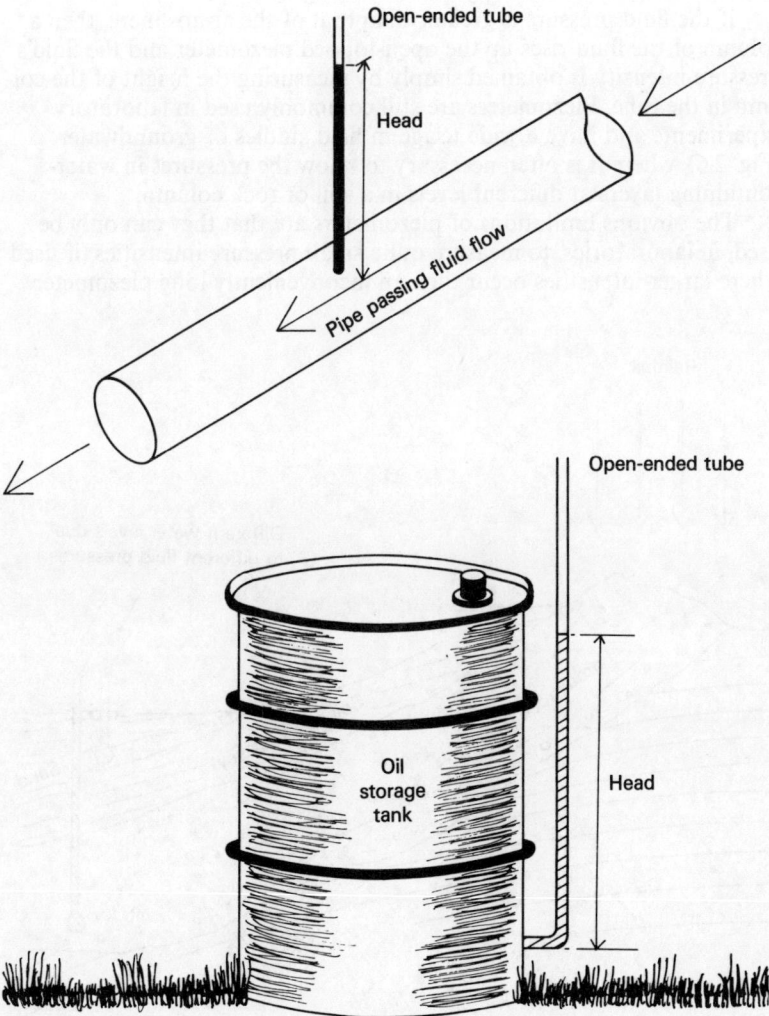

Fig. 2.5 Piezometer installations

2.4 Measuring fluid pressure intensities

Until fairly recently two direct methods of measuring fluid pressure intensities have existed, together with one indirect method. To these have been added a convenient and reliable method arising from modern developments in electronics.

2.4.1 Piezometers

The simplest and oldest of the direct methods is the *piezometer*, which is simply a tube, open at the top to the atmosphere and inserted into the fluid whose pressure intensity is wanted (Fig. 2.5).

If the fluid pressure is greater than that of the atmosphere, then a column of the fluid rises up the open-topped piezometer and the fluid's pressure intensity is obtained simply by measuring the height of the column in the tube. Piezometers are still commonly used in laboratory experiments and have a wide usage in field studies of groundwater (Fig. 2.6), where it is often necessary to know the pressures in water-containing layers at different levels in a soil or rock column.

The obvious limitations of piezometers are that they can only be used, in laboratories, to measure quite small pressure intensities (if used where larger intensities occur then an inconveniently long piezometer

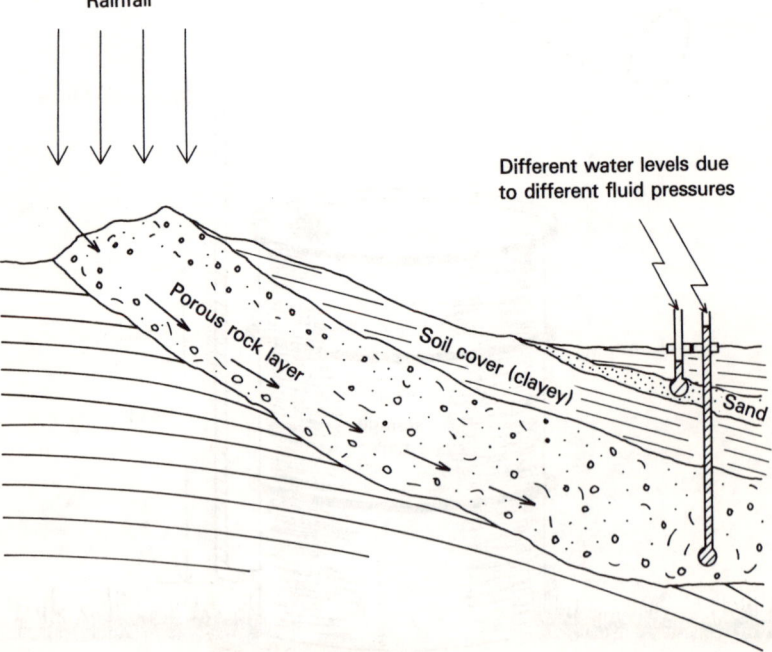

Fig. 2.6 Use of piezometers in groundwater studies

tube would be needed). They cannot be used where fluid pressures lower than atmospheric have to be determined – in such cases the atmospheric pressure on the open end would simply force atmospheric gases into the pressure vessel and probably create a disrupting air lock – and they are fragile and easily broken.

2.4.2 Manometers

To extend the measuring range of the piezometer with an instrument of smaller size and to overcome the problem of lower than atmospheric pressure measurement, the *manometer* was developed. Essentially, a manometer is a piezometer with a U-bend in it. In the bend, a denser fluid (usually mercury, whose density is 13.6 times that of water) is inserted. Since the fluid is so much denser, much greater forces are needed to move it and so much larger pressure intensities can be measured with an instrument of manageable size.

The normal arrangement is as shown in Fig. 2.7 with the contact of the mercury and the fluid, whose pressure is wanted, in the left-hand limb at B. If C is the point on the right-hand limb of the manometer on the same horizontal plane as B, then the pressure at B has to equal that at C.

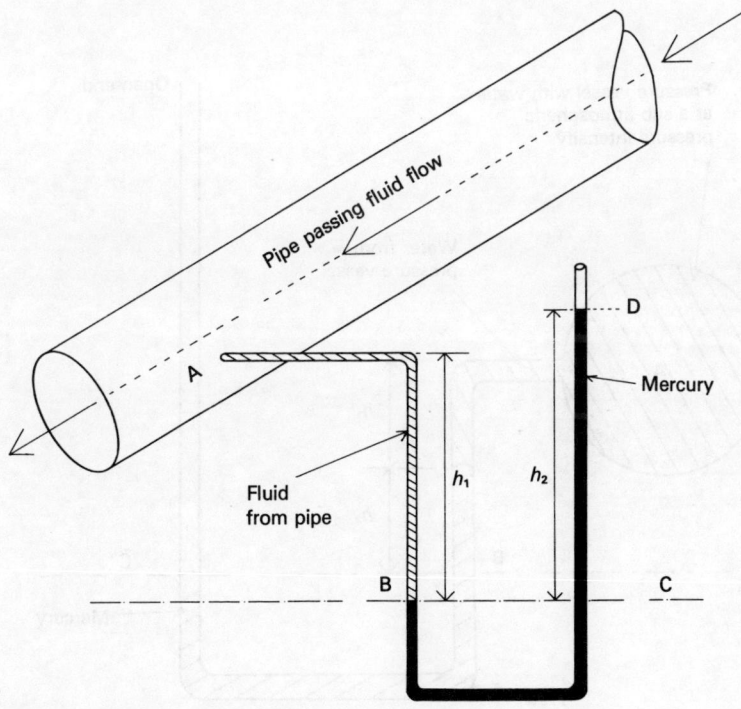

Fig. 2.7 Mercury manometer installed on a pipe

This must be so, since if the pressure intensities were not the same, the two points would have to have different fluid loads on them and would move relative to each other.

The pressure at B is actually of no real interest. We want to know the pressure at A and have to relate it to that at B and C by the pressure intensity concepts we have already developed.

i.e.　　$P_B = P_A +$ the pressure due to the fluid column (h_1) between A
　　　　　　　and B

and　　$P_C = P_D +$ the pressure due to the mercury column (h_2) between C
　　　　　　　and D

since　$P_B = P_C$

$$P_A + (\rho g h_1) \text{ fluid} = P_D + (\rho g h_2) \text{ mercury}$$
$$= 0 + (\rho g h_2) \text{ mercury}$$

since the pressure at D is atmospheric and thus zero in gauge terms

Thus　$P_A = (\rho g h_2)$ mercury $- (\rho g h_1)$ fluid　　　　　　　　[2.5]

If one knows the density of the fluid in the pressure vessel and can measure the lengths of the two fluid columns, the pressure intensity in the vessel can be calculated.

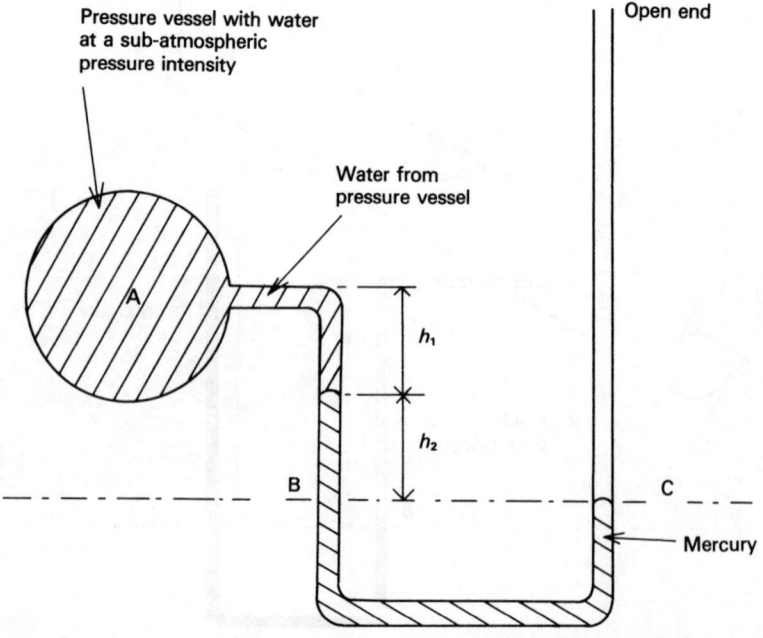

Fig. 2.8 Mercury manometer reading sub-atmospheric pressure intensity

The same approach applies equally well to cases where sub-atmospheric pressure measurement is needed (Fig. 2.8). In this case again, the method is to recognise that the pressure intensity at B has to equal that at C.

$$P_B = P_A + (\rho g h_1) \text{ water} + (\rho g h_2) \text{ mercury}$$
$$P_C = \text{atmosphere value} = \text{gauge zero}$$
$$\therefore \quad P_A = P_B - (\rho g h_1) \text{ water} - (\rho g h_2) \text{ mercury}$$
$$= 0 \quad - (\rho g h_1) \text{ water} - (\rho g h_2) \text{ mercury}$$

and, if both h_1 and h_2 equal 5 cm,

$P_A = -7161.3 \text{ N/m}^2$, a value well below that of the atmosphere.

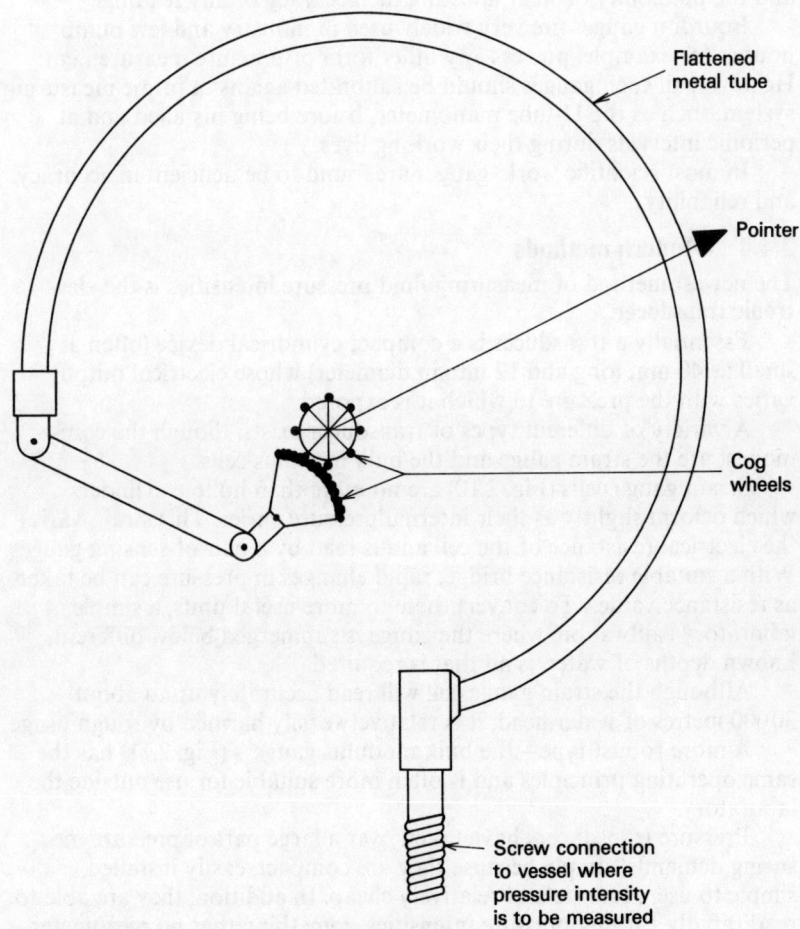

Flattened metal tube

Pointer

Cog wheels

Screw connection to vessel where pressure intensity is to be measured

Fig. 2.9 Bourdon gauge – operating mechanisms

2.4.3 Gauges

The third of the traditional methods is an indirect method, the *Bourdon gauge*, where mechanical factors intervene between the required pressure intensity and its reading (Fig. 2.9).

The heart of the Bourdon gauge is a flattened tube of some material such as phosphor-bronze. As the tube experiences greater pressure it tends to straighten, and conversely, if less pressure occurs, the tube tries to coil up again. These movements, transmitted by a cog system to a pointer, allow a visual gauge reading of the pressure.

Although apparently more convenient (in that none of the calculations needed for manometers are required) the Bourdon gauge, and its various derivatives, are actually less accurate and less reliable. This stems entirely from the various mechanical factors intervening in the reading and the possibility of wear affecting the accuracy of any reading.

Bourdon gauges are very widely used in industry and few pump houses, for example, possess any other form of pressure measurement. However, all such gauges should be calibrated against a prime measuring system, such as the U-tube manometer, before being installed and at periodic intervals during their working lives.

In most scientific work, gauges are found to be deficient in accuracy and reliability.

2.4.4 Modern methods

The newest method of measuring fluid pressure intensities is the **electronic transducer**.

Essentially a transducer is a compact cylindrical device (often as small as 40 mm long and 12 mm in diameter) whose electrical output varies with the pressure to which it is exposed.

A variety of different types of transducer exists, though the commonest are the strain gauge and the bulk modulus cells.

Strain gauge cells (Fig. 2.10) are no more than hollow cylinders which deform slightly as their internal pressure varies. This strain varies the electrical resistance of the cell and is read by a pair of sensing gauges. With a suitable resistance bridge, rapid changes in pressure can be taken as resistance values. To convert these to more useful units, a simple laboratory calibration, where the gauge is submerged below different, known depths of water, is all that is required.

Although the strain gauge cell will read accurately up to about 30 000 metres of water head, it is relatively easily harmed by rough usage.

A more robust type – the bulk modulus gauge – (Fig. 2.11) has the same operating principles and is often more suitable for use outside the laboratory.

Pressure transducers have taken over a large part of pressure measuring demands, simply because they are compact, easily installed, simple to use, accurate and relatively cheap. In addition, they are able to read rapidly varying pressure intensities, something that no piezometer or manometer can hope to achieve.

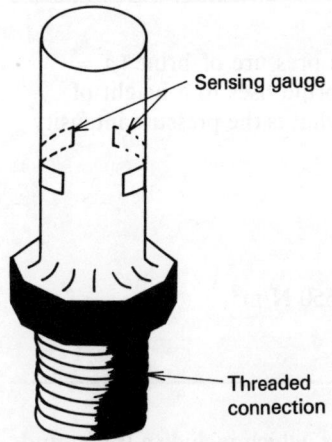

Fig. 2.10 Strain gauge cell

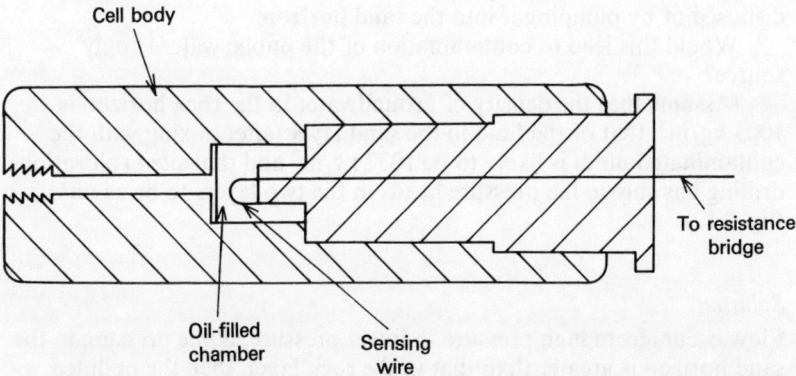

Fig. 2.11 Bulk modulus gauge

Section 2.4 – Self assessment questions

1. Describe with the aid of detailed sketches a piezometer and a U-tube manometer. Explain the operating principles of both pressure intensity measuring instruments.
2. Explain clearly why the U-tube mercury manometer is a more compact instrument than the equivalent piezometer.
3. List the advantages and disadvantages of both the piezometer and the U-tube manometer. Where is each instrument best employed?
4. Detail, with the aid of a sketch, the operating principles of a Bourdon gauge.
5. Why must doubt always exist as to the accuracy of Bourdon gauge readings? How are such readings confirmed in practice?

Example 2.2

A piezometer tube is used to measure the pressure of brine ($\rho = 1030$ kg/m³) flowing in a pipeline. If the brine rises to a height of 0.650 m above the centre of the pipeline, what is the pressure intensity in the pipe?

Solution

The pressure intensity $= \rho \cdot g \cdot h$ N/m²

$$= 1030 \times 9.81 \times 0.650 \text{ N/m}^2$$
$$= 6567.795 \text{ N/m}^2$$

Example 2.3

An area of land is underlain by glacial soils, which include a thick sand horizon. Under the soils is a rock layer. Both the sand and rock layers contain groundwater. The water from the rock horizon is abstracted for a public water supply. It is proposed that contaminated effluent be disposed of by pumping it into the sand horizon.

Would this lead to contamination of the public water supply source?

(Assume that the density of groundwater in the rock horizon is 1003 kg/m³, that of the fluid in the sand layer (after mixing with the contaminated fluid) is likely to be 1037 kg/m³ and that site exploration drilling has shown the pressure heads in the two layers to be as on Fig. 2.12).

Solution

Flow occurs from high pressure to lower pressure. If the pressure in the sand horizon is greater than that in the rock layer, then the polluted water is likely to be carried into the public water supply.

Pressure intensity (sand horizon) $= \rho \cdot g \cdot h$ N/m²

$$= 1037 \times 9.81 \times 4.5 \text{ N/m}^2$$
$$= 45.778 \text{ kN/m}^2$$

Pressure intensity (rock layer) $= 1003 \times 9.81 \times 4.472$ N/m²

$$= 44.000 \text{ kN/m}^2$$

Thus a slightly higher pressure intensity exists in the sand horizon and the potential for the flow of contaminated fluid to the rock aquifer exists.

The proposal is thus potentially dangerous and should not be approved.

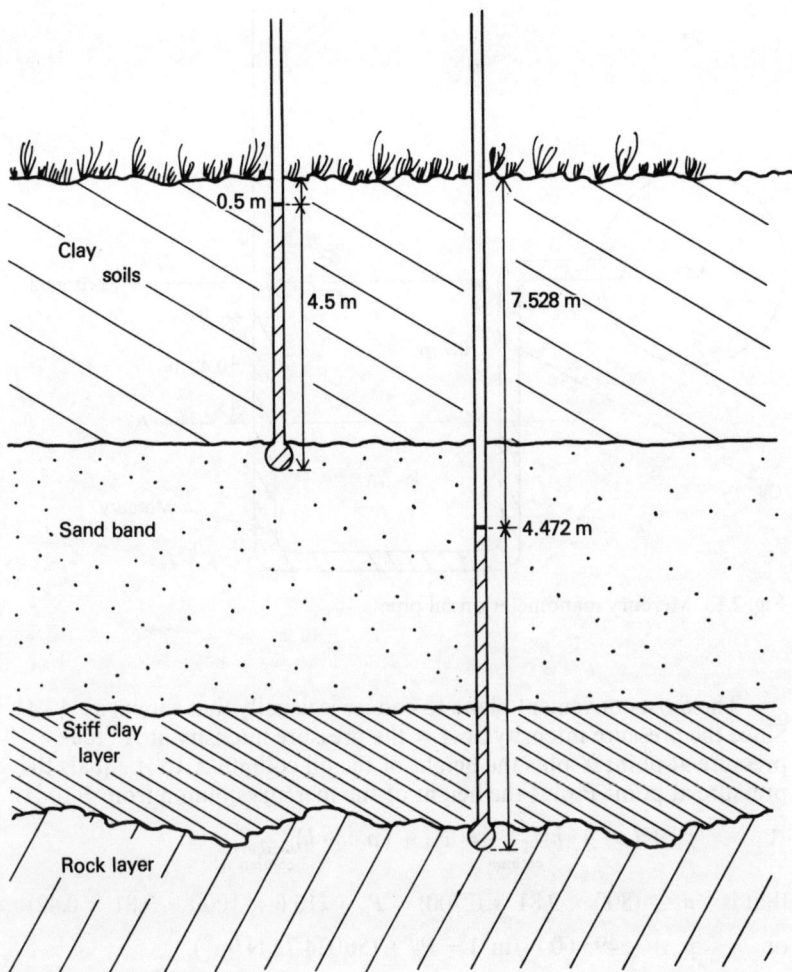

Fig. 2.12 Piezometer installation for liquid waste disposal proposal

---

Example 2.4

The U-tube mercury manometer, installed on a pipe carrying crude oil
($\rho = 800$ kg/m³), gives the readings shown on Fig. 2.13.

Determine the pressure intensity at the centre of the pipe (point x).

Solution

The basic analytical method is always to relate the pressure intensities
in the two limbs of the manometer to a chosen horizontal plane. As the
intensity is uniform on a horizontal plane, this allows the two limbs to
be equated.

24

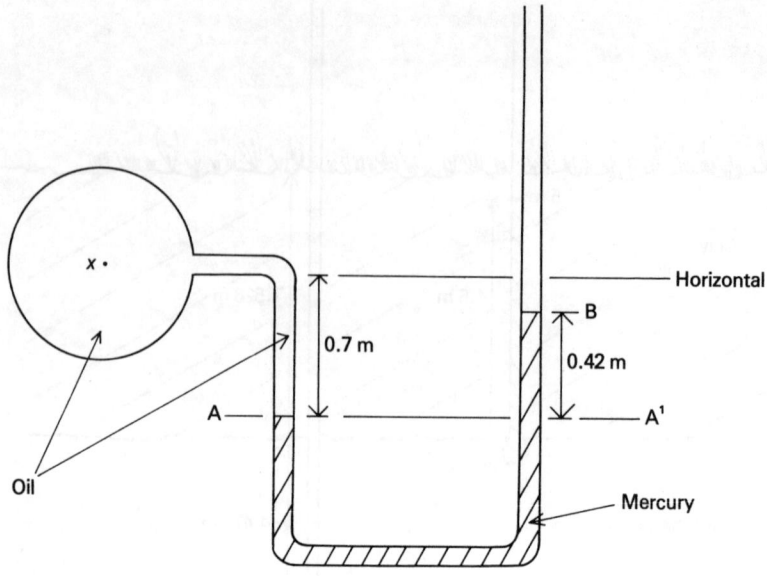

Fig. 2.13 Mercury manometer on oil pipe

The most convenient plane to choose is usually that shown as A–A^1. Since the pressure intensity at A = the pressure intensity at A^1 the pressure at point X plus the height of the oil column X to A equals the pressure at point B plus the height of the mercury column from B to A.

$$\therefore \quad p_x + (\rho \cdot g \cdot h)_{\substack{\text{oil}\\ \text{column}}} = P_B + (\rho \cdot g \cdot h)_{\substack{\text{mercury}\\ \text{column}}}$$

that is $\quad p_x + (800 \times 9.81 \times 0.700) = P_B + (13.6 \times 1000 \times 9.81 \times 0.42)$

or $\quad p_x + (5493.60 \text{ N/m}^2) = P_B + (56034.72 \text{ N/m}^2)$

and, since P_B = atmospheric pressure = zero in gauge terms,

$p_x = 56034.72 - 5493.60$

$\underline{ = 50541.12 \text{ N/m}^2}$

The pipeline thus contains fluid at a pressure significantly greater than that of the atmosphere. It will be noted that the mercury level stands much higher in the right-hand manometer limb than it does in the other limb.

Example 2.5
If, under different operating conditions the same manometer, as in Example 2.4, has the readings shown below, what is the pressure intensity?

Left-hand manometer limb — oil/mercury interface at 0.3 m below centre line of pipe

Right-hand manometer limb — mercury surface at 0.7 m below centre line of pipe

Solution
As before, a suitable horizontal plane has to be chosen to allow analysis. The easiest plane to use is that through the mercury surface (A^1) in the right-hand limb of the manometer.

If this is taken then the pressure at the equivalent point (A) on the left-hand limb (P_A) is

$$P_A = P_x + (\rho \cdot g \cdot h)_{\substack{\text{oil column} \\ \text{above A}}} + (\rho \cdot g \cdot h)_{\substack{\text{mercury column} \\ \text{above A}}}$$

(where P_x is again the pressure intensity at the pipe's centre)

$$= P_x + (800 \times 9.81 \times 0.3) + (13.6 \times 1000 \times 9.81 \times 0.4)$$

$$= P_x + (2354.4 \text{ N/m}^2) + (53\,366.4 \text{ N/m}^2)$$

$$= P_{A^1}$$

$$= 0$$

(since point A^1 on the right-hand limb is under only atmospheric pressure)

$$\therefore \quad P_x = -55\,720.8 \text{ N/m}^2$$

and the pressure intensity at the centre of the pipeline is under a much less than atmospheric pressure intensity.

In this case it is noticeable that the mercury level is higher in the left-hand limb of the manometer.

Example 2.6
A Bourdon gauge, calibrated to read pressures in metres of water, is installed in a pipeline (Fig. 2.14). After some months of use, the gauge reads 2.370 m of water. To determine the accuracy of the gauge, a U-tube mercury manometer is screwed into a fitting on the same plane as the gauge. The manometer shows the following readings:

Left-hand limb	Right-hand limb
Water to 0.350 m below pipe centre Mercury at 0.350 m below pipe centre	Mercury level at 0.150 m below pipe centre
Is the gauge accurate?	

Solution
Analysing the U-tube results – so –

$$P_A = P_{A^1}$$

$$\therefore \quad P_x + (\rho \cdot g \cdot h) \quad \text{water column} = P_B + (\rho \cdot g \cdot h) \quad \text{mercury column}$$

i.e. $\quad P_x + (1000 \times 9.81 \times 0.35) = 0 + (13.6 \times 1000 \times 9.81 \times 0.2)$

or $\quad\quad P_x + (3433.5 \text{ N/m}^2) = 0 + (26\,683.2 \text{ N/m}^2)$

$$\therefore \quad\quad\quad\quad\quad P_x = 23\,249.7 \text{ N/m}^2$$

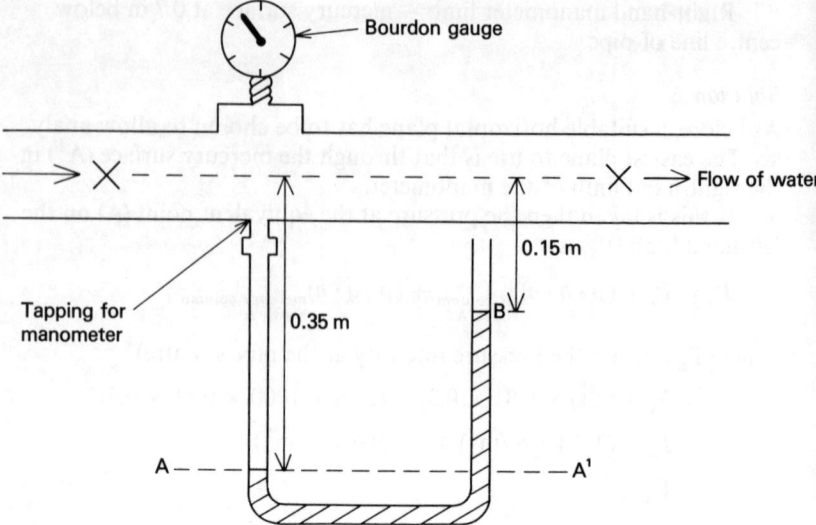

Fig. 2.14 Calibration of a Bourdon gauge

and thus is equivalent to a water column of height (h) metres,

where $h = \dfrac{P}{\rho \cdot g} = \dfrac{23\,249.7}{1000 \times 9.81}$

$\qquad\qquad = 2.37$ metres of water

Thus Bourdon gauge is still reading accurately.

Chapter 3

Pressure forces on surfaces

To design fluid-retaining tanks, walls and vessels, engineers have to know the forces that the fluid exerts and the locations of these forces. A knowledge of pressure intensities alone is simply not adequate, as has already been indicated in sections 2.2. and 2.3.

The simplest examples to consider are those where the fluid forces act on planar surfaces. Two distinct cases, of course, exist – where the planar surface is everywhere exposed to the same pressure intensity, and where the plane suffers an increasing pressure intensity, usually varying with depth.

3.1 Uniform pressure intensities

The first case of uniform pressure intensities presupposes that the planar body lies parallel with the surface of the fluid.

The pressure intensity is, as before,

$p = \rho \cdot g \cdot h$ newtons/metre^2

and the fluid force on any elementary area (a) is

$F = p \cdot a = \rho \cdot g \cdot h \cdot a \cdot$ newtons   [2.4]

Because these elementary forces are uniform over the entire area (A), their resultant has to act through the centre of gravity of the body. In this particular case, the *centre of pressure* therefore coincides with the *centre of gravity*.

The uplift experienced by foundations, constructed below the groundwater level (section 2.3), is one example of uniform pressure intensities on a planar structure.

3.2 Varying pressure intensities

The second case – that of varying pressure intensity on a planar body – is somewhat more complex. However, it is of much greater practical application and thus worthy of detailed analysis. The great majority of fluid-retaining structures have to withstand pressure intensities which increase with fluid depth down the structure.

The simplest case to imagine is of a wall with hydrostatic pressures on only one side (Fig. 3.1). Along any narrow horizontal strip (M–N), the pressure intensity is uniform and has a value of

$$p = \rho \cdot g \cdot h \, \text{N/m}^2$$

and the pressure force on this elementary strip is

$$F = (\rho \cdot g \cdot h) \, (\text{area of elementary strip})$$
$$= (\rho \cdot g \cdot h) \, (y \cdot dh) \, \text{newtons}.$$

To obtain the total force (F) over the wall it is necessary to integrate the above expression over the total height of the wall – so:

$$F = \int_B^0 \rho \cdot g \cdot h \cdot y \cdot dh$$
$$= \rho \cdot g \int_B^0 y \, dh \cdot h$$

As the term to be integrated is the **moment** of the area of strip M–N about the fluid surface, its integral must be the total area A, times the distance from the fluid surface to the centre of gravity ($\bar{h}$) of the wall.

i.e. $\quad F = \rho \cdot g \cdot A \cdot \bar{h}$ [3.1]

Knowing the numerical value of the force is, however, not enough. The location of the force must be determined by taking moments about the fluid's surface.

The moment of the force on the elementary strip M–N is

$$M = \begin{pmatrix} \text{the force on} \\ \text{the strip} \end{pmatrix} \times \begin{pmatrix} \text{the lever arm distance to the} \\ \text{fluid's surface} \end{pmatrix}$$
$$= (\rho \cdot g \cdot h \cdot y \cdot dh) \times (h)$$

and the total moment (M) is the integral of this.

i.e. $\quad M = \rho \cdot g \int_B^0 y \cdot h^2 \cdot dh$

29

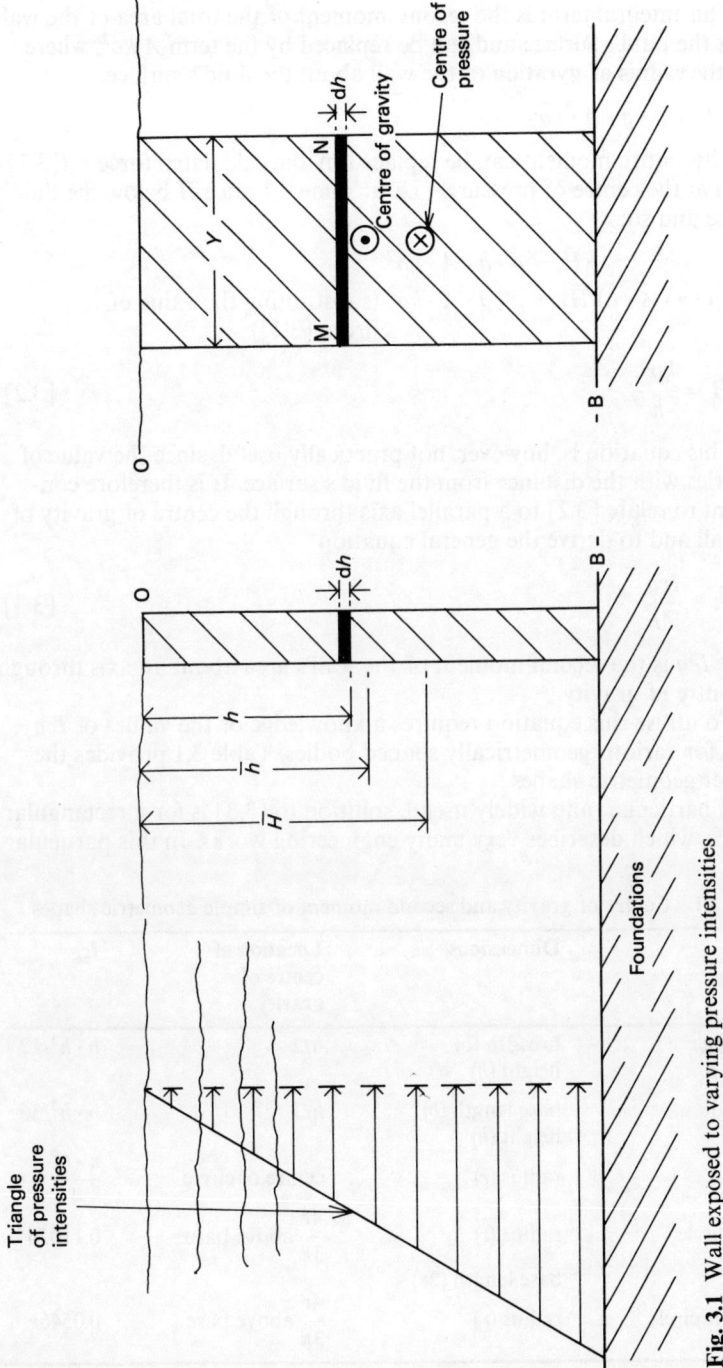

Fig. 3.1 Wall exposed to varying pressure intensities

This integral term is the second moment of the total area of the wall about the fluid's surface and can be replaced by the term $A\,ko^2$, where ko is the **radius of gyration** of the wall about the fluid's surface.

$$\therefore \quad M = \rho \cdot g \cdot A \cdot ko^2$$

This entire moment can be replaced by the calculated force F ([3.1]) acting at the centre of pressure – i.e. at some distance $\bar{H}$ below the fluid's surface and so:

$$F\bar{H} = \rho \cdot g \cdot A \cdot ko^2$$

or $(\rho \cdot g \cdot A \cdot \bar{h})(\bar{H}) = \rho \cdot g \cdot A \cdot ko^2$ (substituting the value of F from [3.1])

$$\therefore \quad \bar{H} = \frac{ko^2}{\bar{h}} \tag{3.2}$$

This equation is, however, not practically useful, since the value of ko varies with the distance from the fluid's surface. It is therefore convenient to relate [3.2] to a parallel axis through the centre of gravity of the wall and to derive the general equation

$$\bar{H} = \bar{h} + \frac{Icg}{A\bar{h}} \tag{3.3}$$

where Icg is the second moment of the wall's area about an axis through the centre of gravity.

To utilise this equation requires a knowledge of the values of Icg and $\bar{h}$ for various geometrically shaped bodies. Table 3.1 provides the simpler geometric shapes.

A particular, and widely useful, solution to [3.3] is for a rectangular surface, which describes very many engineering works. In this particular

Table 3.1 Centre of gravity and second moment of simple geometric shapes

Shape	Dimensions	Location of centre of gravity	Icg
Rectangle	breadth (b) height (h)	$h/2$	$b \cdot h^3/12$
Triangle	base length (b) height (h)	$h/3$	$b \cdot h^3/36$
Circle	radius (r)	centre of circle	$\dfrac{\pi \cdot r^4}{4}$
Semi-circle	radius (r)	$\dfrac{4r}{3\pi}$ above base	$0.1102\,r^4$
	base length ($2r$)		
Quarter circle	radius (r)	$\dfrac{4r}{3\pi}$ above base	$0.0546r$

case

$$\bar{H} = 2/3\,h \qquad\qquad\qquad [3.4]$$

This is, of course, appropriate for the analysis of horizontal fluid thrust on a dam wall (section 2.2) and seepage uplift forces under a dam (section 2.3), as in both cases the surface exposed to the fluid force was defined as rectangular.

3.3 Pressure forces on non-planar surfaces

More complex cases are where the surface withstanding the fluid pressure force is not a plane (Fig. 3.2).

In such cases, it is necessary to resolve the pressure intensities along chosen axes, at right angles. The areas on which the forces act are also projected onto the chosen axes.

Self assessment questions

1. Does the term 'centre of gravity' mean the same as the term 'centre of pressure'?
2. Can the centre of pressure coincide with the centre of gravity?
3. In all other cases is the centre of pressure higher or lower than the centre of gravity?

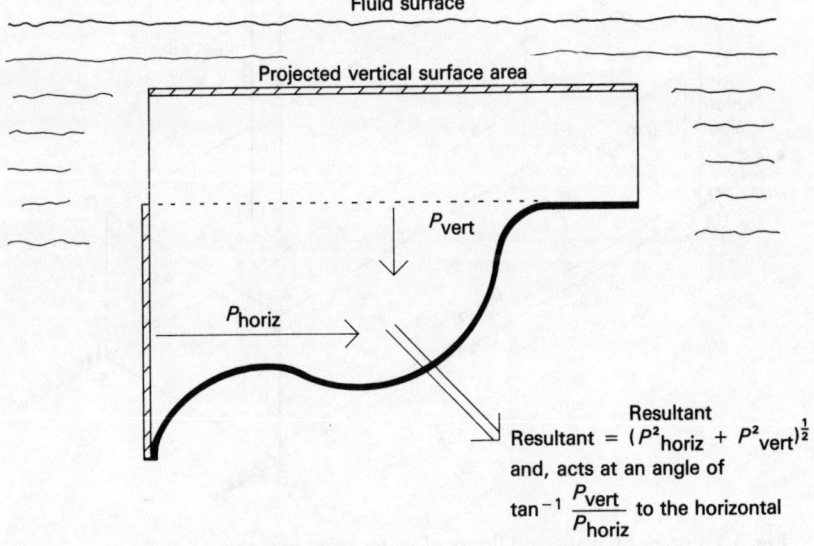

Fig. 3.2 Resolving pressure forces on a non-planar surface

4. Does the magnitude of the total pressure force on a body, subjected to differing pressure intensities, depend on the value of the centre of gravity $\bar{h}$?

5. Explain why the magnitude of the total pressure force is dependent on the value of the centre of gravity and yet the force acts at a lower point (the centre of pressure).

6. Does the centre of pressure always lie at two-thirds of the submerged depth of a body, subjected to differing pressure intensities?

7. Where a surface is not a plane, is it necessary to project the area of the surface on to mutually perpendicular axes?

8. In cases such as Q. 7, can any axes be chosen?

Examples

Example 3.1

A rectangular storage tank is 6 m long, 2 m high and 2.5 m wide and is filled by a vertical pipe through the top of the tank (Fig. 3.3). The tank is designed to remain full in all normal operations. Determine the magnitude and locations of the pressure forces on the floor and end sides of the tank, if the tank is full up to the level C–D.

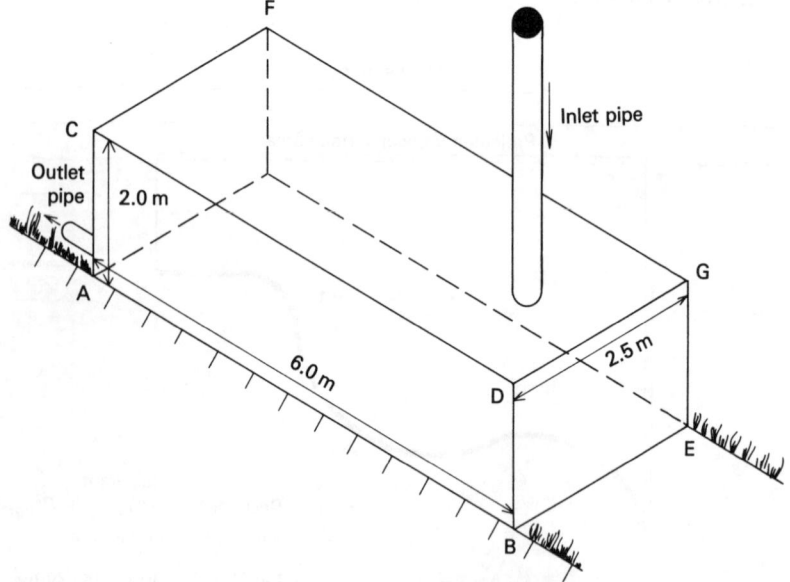

Fig. 3.3 Forces on walls and floors of water storage tank

Solution
(a) Force on tank floor. The fluid pressure intensity is uniform over the entire floor of the tank and thus the pressure force is

$F = \text{p.a.} = (\rho \cdot g \cdot h)(a)$
$= (1000 \times 9.81 \times 2)(6 \times 2.5)$ newtons
$= 294\,300$ newtons
$= 294.300$ kN

This force is uniformly distributed and its resultant can be considered to pass through the centre of gravity of the tank's floor.

(b) Force on end sides of tank. The centre of gravity of both end sides lies 1 m below the tank's top surface. The pressure force is

$$F = \rho \cdot g \cdot \bar{h}\,(A) \hspace{4cm} [3.1]$$
$$= 1000 \times 9.81 \times 1(2 \times 2.5)$$
$$= 49.05 \text{ kN}$$

This force is located at a point two-thirds of the distance from the water level in the tank ([3.4]) i.e. at 1.33 m from the top of the tank.

Example 3.2
Due to a blockage of the outlet pipe, from the same water storage tank, water rises in the inlet pipe and stands 3.7 m above the top of the tank. Determine the magnitude and locations of the pressure forces on the floor and end side (C–A) of the tank.

Solution
(a) Force on tank floor. As the tank floor is again horizontal, the pressure intensity must be the same over the entire floor area.

as before $\quad F = \text{p.a.} = (\rho \cdot g \cdot h)(a)$
$$= (1000 \times 9.81 \times 5.7)(6 \times 2.5)$$
$$= 838.755 \text{ kN}$$

also, as before, this force can be taken as acting through the centre of the floor's area.

(b) Force on the end walls (C–A). The centre of gravity of side C–A lies 4.7 m below the water level in the inlet pipe.

$\therefore \quad F = \rho \cdot g \cdot \bar{h} \cdot A = 1000 \times 9.81 \times 4.7 \times 2.0 \times 2.5$
$$= 230.535 \text{ kN}$$

This acts at the centre of pressure $(\bar{H})$ of the face C–A

where $\quad \bar{H} = \bar{h} + \dfrac{Icg}{A\bar{h}}$

$$= 4.7 + \dfrac{(2.5)(2^3)}{12} \Big/ (2 \times 2.5)/4.7$$

$= 4.771$ m below the water surface in the inlet pipe

Example 3.3

A lock gate in a canal is 5 m wide. When barges use the lock, a water depth of 7.5 m occurs on the upstream side of the gate, and only 3.0 m on the downstream face (Fig. 3.4).

Determine the magnitude and location of the resultant force on the gate.

Solution

Upstream face.

Area of gate exposed to fluid force $= 5 \times 7.5$ m^2

$$= 37.5 \text{ m}^2$$

Depth to centre of gravity $(\bar{h}) = \frac{1}{2}$ wetted depth

$$= \tfrac{1}{2}(7.5)$$

$$= 3.75$$

Fluid force $= \rho \cdot g \cdot \bar{h} \cdot A$ newtons

$= 1000 \times 9.81 \times 3.75 \times 37.5$

$= 1379.531$ kN, located at 2.5 m above the base of the gate.

Downstream face.
Similarly,

Area of gate exposed to fluid force $= 5 \times 3$

$$= 15 \text{ m}^2$$

Depth to centre of gravity $(\bar{h}) = 1.5$ m

Fluid force $= \rho \cdot g \cdot \bar{h} \cdot A$ newtons

$= 1000 \times 9.81 \times 1.5 \times 15$

$= 220.725$ kN,   located at 1 m above the base of the gate.

The difference of these two forces is 1158.806 kN

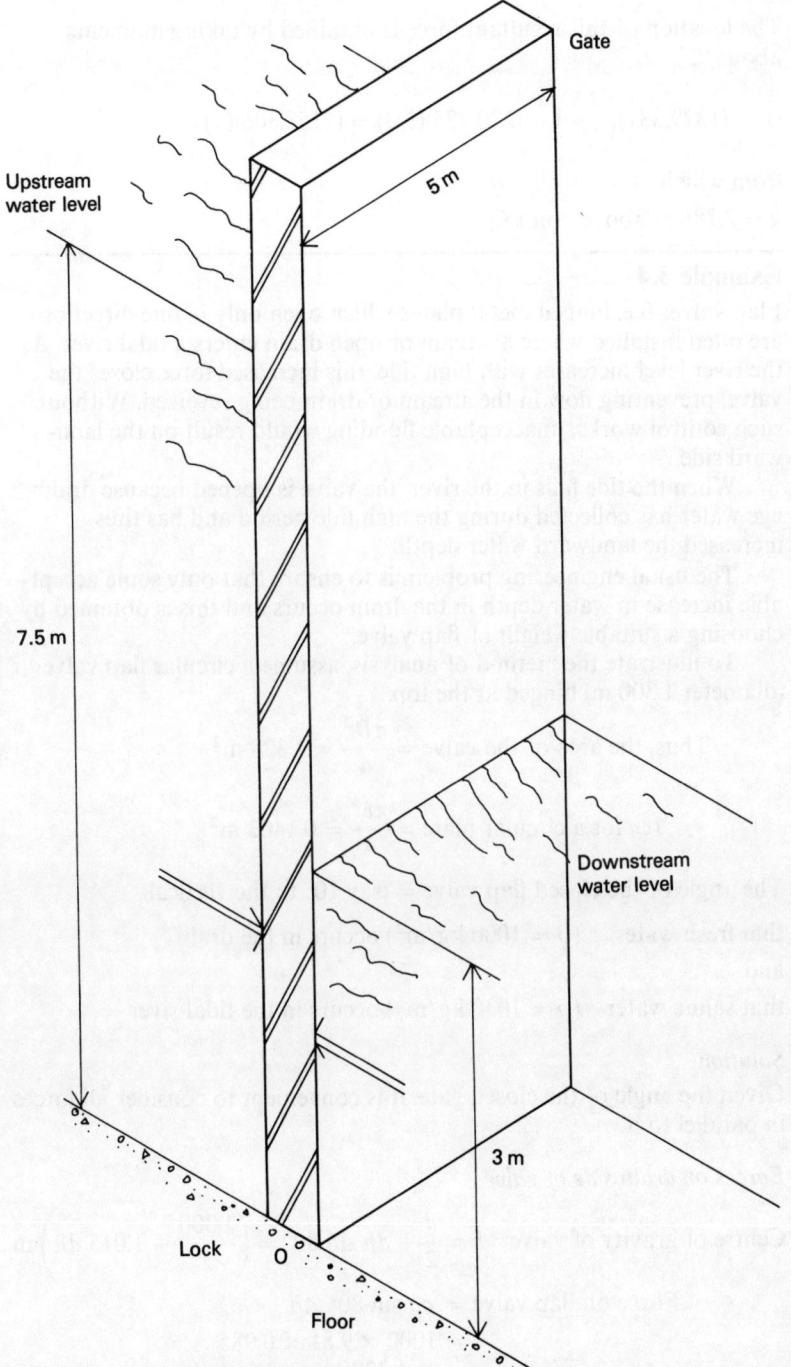

Fig. 3.4 Forces on lock gate

The location of this resultant force is obtained by taking moments about 'O'.

i.e. $(1379.531)\left(\dfrac{7.5}{3}\right) - (220.725)(3/3) = (1158.806)(y)$

from which

$y = 2.786$ m above point O.

Example 3.4

Flap valves (i.e. hinged metal plates which open only in one direction) are often installed where a stream or open drain enters a tidal river. As the river level increases with high tide, this increased force closes the valve, preventing flow in the stream or drain being reversed. Without such control works, unacceptable flooding would result on the landward side.

When the tide falls in the river, the valve is opened because drainage water has collected during the high tide period and has thus increased the landward water depth.

The usual engineering problem is to ensure that only some acceptable increase in water depth in the drain occurs and this is obtained by choosing a suitable weight of flap valve.

To illustrate the method of analysis, assume a circular flap valve (diameter 1.300 m) hinged at the top.

$$\text{Thus, the area of the valve} = \frac{\pi D^2}{4} = 1.328 \text{ m}^2$$

$$Icg \text{ for a circular plate} = \frac{\pi r^4}{4} = 0.1403 \text{ m}^2$$

The angle of the closed flap valve = (say 10° to the vertical

that fresh water   ($\rho = 1000$ kg/m^2) occurs in the drain
and
that saline water   ($\rho = 1030$ kg/m^3) occurs in the tidal river

Solution

Given the angle of the closed gate, it is convenient to consider all forces in parallel to it.

Forces on drain side of valve

Centre of gravity of valve $\bar{h} = \dfrac{d}{2} + dh \sin 80° = \left(\dfrac{1.300}{2} + 1.015\,dh\right)$ m

$$\begin{aligned}\text{Force on flap valve} &= \rho g \sin 80°\, A\bar{h} \\ &= 1000 \times 9.81 \times 0.985 \\ &\quad \times 1.328(\bar{h}) \text{ newtons}\end{aligned}$$

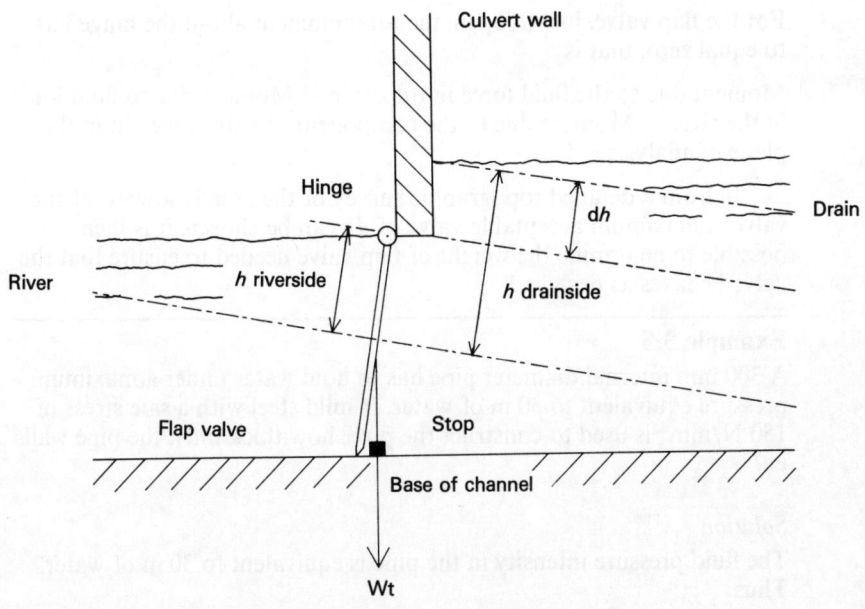

(Low tide condition - flap valve about to open)

Fig. 3.5 Flap valve controlling drainage to tidal river

Location of this force is $\bar{H} = \bar{h} + \dfrac{Icg}{A\bar{h}}$

$$= \bar{h} + \frac{0.1403}{1.328(\bar{h})} \text{ m from the hinge}$$

Forces on river side of valve

Centre of gravity of valve is $\bar{h} = \dfrac{d}{2} = 0.650$ m

Force of river on gate $= \rho g \sin 80° \, A\bar{h}$

$$= 1030 \times 9.81 \times 0.985$$
$$\times 1.328 \times 0.65 \text{ newtons}$$
$$= 8591.20 \text{ newtons}$$

Location of this force is $\bar{H}$ $= h + \dfrac{Icg}{A\bar{h}}$

$$= 0.65 + \frac{0.1403}{1.328} \times 0.65$$

$$= 0.813 \text{ m from the hinge.}$$

For the flap valve, just to open, the total moment about the hinge has to equal zero, that is

Moment due to the fluid force in the drain — Moment due to fluid force in the river — Moment due to the component of valve's weight in the plane of analysis = 0.

If, from a detailed topographic survey of the area landward of the valve, a maximum acceptable valve of dh can be chosen, it is then possible to determine the weight of flap valve needed to ensure that the valve behaves as designed.

Example 3.5

A 300 mm internal diameter pipe has to hold water under a maximum pressure equivalent to 30 m of water. If mild steel with a safe stress of 150 N/mm^2 is used to construct the pipe, how thick must the pipe walls be?

Solution

The fluid pressure intensity in the pipe is equivalent to 30 m of water. Thus

$$P = \rho gh = 1000 \times 9.81 \times 30$$
$$= 294\ 300 \text{ N/m}^2$$
$$= 0.294 \text{ N/mm}^2$$

since one million millimetre^2 (10^6 mm^2) are equal to one metre^2 (m^2).

The bursting of a pipe can be thought of as a tendency for the top half to separate from the bottom half. The only force acting against this tendency is the hoop tension (T) of the pipe walls. If the pipe is to have the thinnest possible walls then the bursting pressure force must exactly equal the hoop tension.

Since the pipe walls are not plane it is necessary to project the area of these, exposed to the calculated pressure intensity, on to a suitable axis. The most convenient axis is the horizontal plane separating the upper and lower halves of the pipe.

If only a 1-mm length of the pipe is considered (a valid limitation since each square millimetre of the pipe wall has to achieve the necessary minimum strength) then the projected area exposed to the fluid pressure intensities is rectangular and has sides of

$2r$ millimetres and 1 millimetre = $2r$ mm^2

Further, the fluid pressure intensities are obviously symmetrically disposed about the pipe wall and their resultant can be taken as a single force acting at right angles to the plane of area projection.

Thus the total bursting pressure = $P \times 2r \times 1$

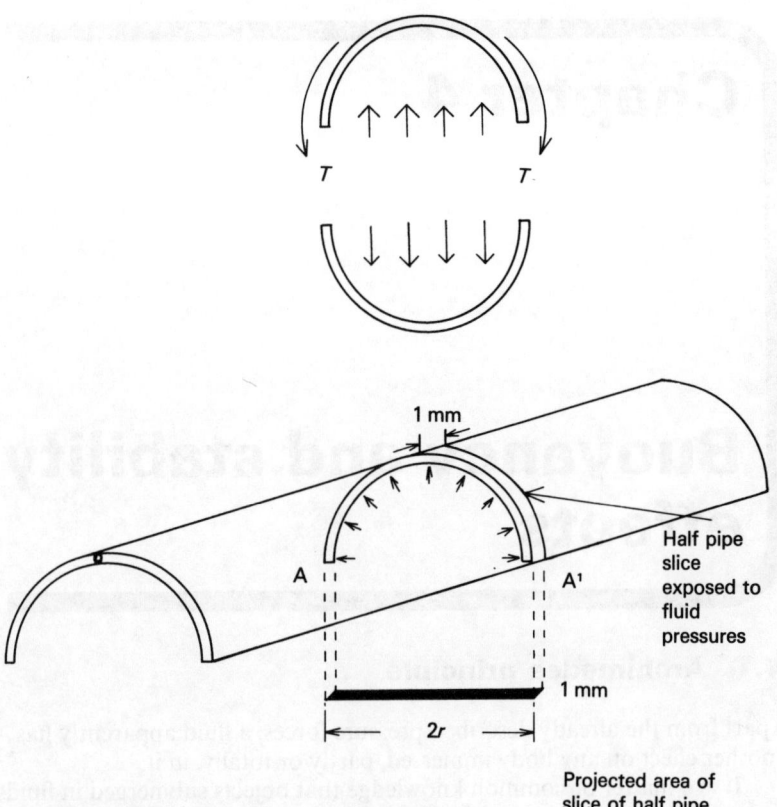

Fig. 3.6 Internal forces on walls of a pipe

which has to equal $2T$

$\therefore \quad T = $ r.p. newtons.

The hoop tension equals the safe stress multiplied by the area over which the stress acts. This particular area equals the pipe wall thickness (t mm) times the length of pipe being analysed (1 mm).

$\therefore \quad T = $ t.p. = safe stress × wall thickness

or wall thickness $(t) = \dfrac{\text{t.p}}{\text{s.s}} = \dfrac{150 \text{ mm} \times 0.294}{150 \text{ N/mm}^2} \text{ N/mm}^2$

$\qquad\qquad\qquad = \underline{0.294 \text{ mm}}$

Thus the minimum necessary wall thickness for this pipe is only 0.294 mm. In practice a stronger, and thus a thicker, pipe wall is needed to allow for the forces due to handling and laying.

Chapter 4

Buoyancy and stability effects

4.1 Archimedes principle

Apart from the already described pressure forces, a fluid apparently has another effect on any body immersed, partly or totally, in it.

It is a matter of common knowledge that objects submerged in fluids appear, for some reason, to lose part of their weight. For example, a young girl in a swimming pool can tow a fully grown man, whom she would be unable to drag along on dry land.

This effect is easily demonstrated by a precise experiment. Take a house brick, or any other convenient regular object, measure its volume and then its weight in air. If the brick is found to have a weight of 5 kg and a volume of 1875 cm^3, its weight when fully submerged in water will fall to only 3.125 kg.

The apparent loss of 1875 grams of weight is easily explained as due to the difference in fluid forces on the upper and lower surfaces of the brick (Fig. 4.1).

Prior to the brick being placed in the fluid the downward force on the area represented by CD was equal to the weight of the fluid column it supported. As the fluid was in equilibrium, the downward force was exactly balanced by an equal upward force. When the brick is immersed to occupy the volume represented by ABCD, the fluid force on its upper surface is the weight of the fluid column above the area represented by AB. However, since the fluid is still undisturbed below the plane through C–D, all fluid pressure intensities on this plane are as before and the same upward balancing force will continue to act – despite the immersion

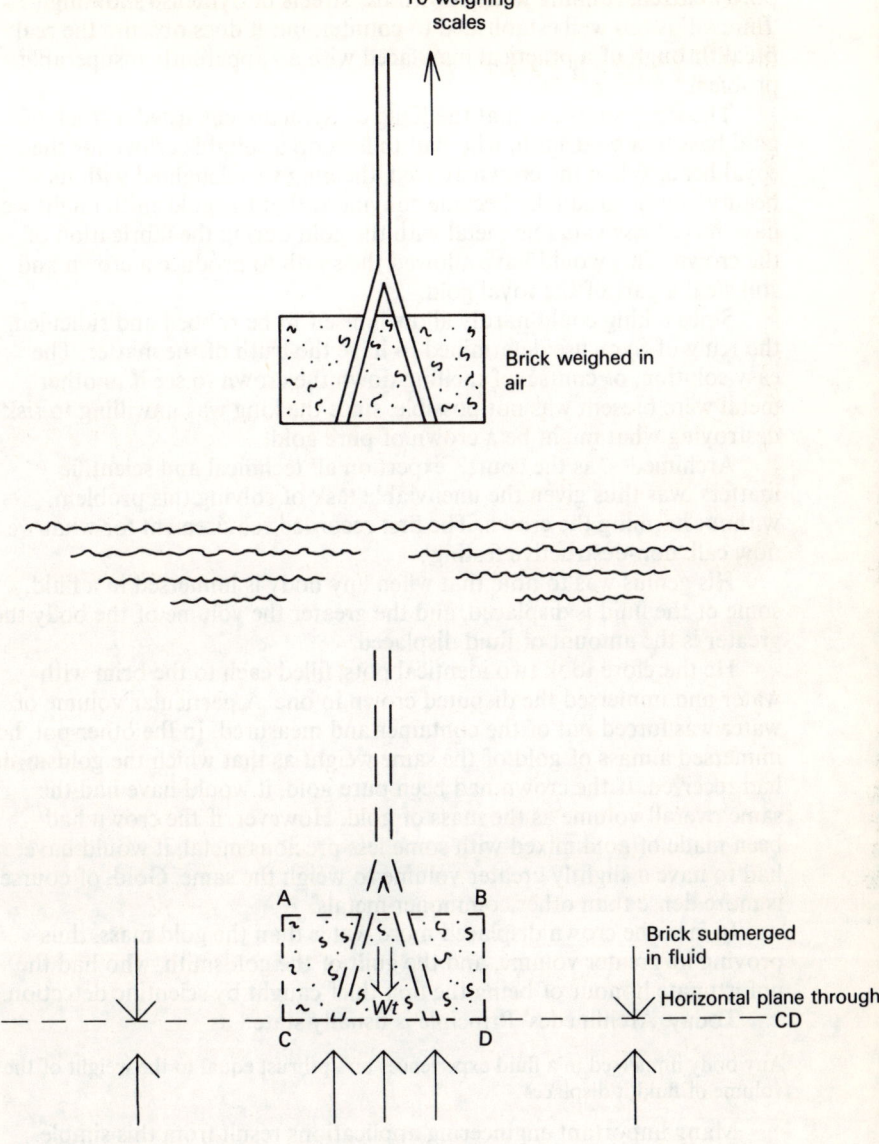

Fig. 4.1 Determining the buoyancy force on a brick

of the house brick. Thus the weight of the brick vertically downwards will be counteracted by an upward force equal to the weight of the volume of fluid displaced by the immersion of the solid body.

This principle was discovered by the Greek scientist Archimedes, some 2200 years ago. The popular image of Archimedes, as a naked portly citizen, running wildly down the streets of Syracuse shouting 'Eureka!' is too well established to counter, but it does obscure the real breakthrough of a practical man faced with an apparently insuperable problem.

The story simply is that the King of Syracuse entrusted a stock of gold bars to a goldsmith, who had to fashion a suitable crown for the royal head. When the crown arrived, the king was delighted with its beauty, but he gradually became convinced that the goldsmith might well have mixed less valuable metal with the gold during the fabrication of the crown. This would have allowed the smith to produce a crown and still steal a part of the royal gold.

Since a king could hardly allow himself to be robbed and ridiculed, the King of Syracuse determined to have the truth of the matter. The easy solution, of course, of melting down the crown to see if another metal were present was not possible, since the king was unwilling to risk destroying what might be a crown of pure gold.

Archimedes, as the court's expert on all technical and scientific matters, was thus given the unenviable task of solving this problem, without harming the crown. The first recorded requirement for what we now call 'non-destructive testing'.

His genius was to note that when any body is immersed in a fluid, some of the fluid is displaced, and the greater the volume of the body the greater is the amount of fluid displaced.

He therefore took two identical pots, filled each to the brim with water and immersed the disputed crown in one. A particular volume of water was forced out of the container and measured. In the other pot, he immersed a mass of gold of the same weight as that which the goldsmith had received. If the crown had been pure gold, it would have had the same overall volume as the mass of gold. However, if the crown had been made of gold mixed with some less precious metal, it would have had to have a slightly greater volume to weigh the same. Gold, of course, is more dense than other commoner metals.

In fact, the crown displaced more water than the gold mass, thus proving its greater volume, and the guilt of the goldsmith, who had the unfortunate honour of being the first thief caught by scientific detection.

Today, Archimedes' Principle is usually stated as

Any body immersed in a fluid experiences an upthrust equal to the weight of the volume of fluid it displaces.

Many important engineering applications result from this simple fact.

A common problem is of pipelines which have to be laid under water at river, lake and sea crossings. Such pipes would be damaged at their joints if they tended to float in some conditions and to sink in others. Yet pipes are at times full of the water, or whatever other fluid they have to transmit, and are, at other times, emptied for maintenance.

Because of the differing operational conditions which may occur in the life of the pipe, it is imperative to check the buoyancy forces for the pipe full and the pipe empty case.

Self assessment questions

1. How would you determine the volume of the brick used in the Archimedes experiment? Assume that it is irregular enough to make measurement with a ruler inaccurate.
2. Explain clearly how the weight of a body in air is reduced when the body is submerged (totally or partly) in a fluid.
3. Clearly define buoyancy force and explain it in terms of the difference in fluid pressure forces on the lower and upper surfaces on the body.
4. Why should we consider the buoyancy effects, on a submerged pipeline, in other than the case of the pipe full of the fluid it is conveying?
5. Clearly distinguish between stable, unstable and neutral equilibrium conditions for a floating body.
6. Until recently, buoyancy to most civil engineers was an effect limited to cases where structures had to be laid in river beds or to buildings founded in waterlogged soils beside rivers. Now a wider use of buoyancy is necessary. Why is this?

Examples

Example 4.1

A pipeline has to be laid below the surface of a lake. The pipe is 300 mm in internal diameter, its walls are 12.5 mm thick and its weight (per metre length) is 150 newtons. Will buoyancy effects create any problems?

Solution

Consider the buoyancy forces both for the pipe full and the pipe emptied cases.

Pipe full case. Here the forces acting on the pipe are:

Weight of pipe + weight of contained − weight of volume of water
                    water   displaced by pipe

              ↓   ↓   ↑

(note the directions of the arrows indicate the directions in which the forces act on the pipe) and for a 1-metre length of pipe this

$$= 150 \text{ N} + \begin{pmatrix} \text{volume of contained} \\ \text{water times its} \\ \text{density and the} \\ \text{gravitational force} \end{pmatrix} - \begin{pmatrix} \text{external volume of the pipe} \\ \text{times the density of the water} \\ \text{it displaces times the} \\ \text{gravitational force} \end{pmatrix}$$

$$= 150 \text{ N} + \left(\frac{\pi d^2}{4} \times 1 \times \rho \times g \right) - \left(\frac{\pi D^2}{4} \times 1 \times \rho \times g \right)$$

where d = the internal diameter of the pipe (0.3 m)

and D = the external diameter of the pipe (0.325 m)

$$= 150 \text{ N} + \left(\frac{22}{7} \times \frac{0.3^2}{4} \times 1 \times 1000 \times 9.81 \right)$$

$$- \left(\frac{22}{7} \times \frac{0.325^2}{4} \times 1 \times 1000 \times 9.81 \right)$$

$$= 150 \text{ N} + 693.707 \text{ N} - 814.142 \text{ N}$$

$$= 29.565 \text{ N per metre length, acting downwards.}$$

Thus when the pipe is full of water, its remnant weight still acts to prevent the pipe floating or rising.

Pipe empty case. In this case, however, the only forces which act are

$$\begin{matrix} \text{the weight of} \\ \text{the empty pipe} \\ \downarrow \\ = 150 \text{ N} \end{matrix} \quad \begin{matrix} \_ \\ \end{matrix} \quad \begin{matrix} \text{the weight of the volume} \\ \text{of displaced water} \\ \downarrow \\ - 814.142 \text{ N} \end{matrix}$$

$$= 664.142 \text{ N per metre length of pipe}$$

and here a large force will tend to make the pipe rise up in the water and thus will strain and finally break the pipe joints.

In such cases the pipe has to be anchored to its bed (by strapping the pipe to piles into the bed or by bolting the pipe to the bed, if it is rock or other suitably solid material), or has to be weighed down (with, for example a concrete cover of suitable weight). In either case, the tensions in the straps and bolts or the weight of concrete needed can simply be obtained from the above calculations.

Buoyancy forces affect structures other than pipelines immersed in water. Another common example (referred to in section 2.3) is where foundations have to be constructed below the groundwater level (Fig. 4.2). In this case, the structure displaces a volume of the groundwater, and suffers a partial reduction in structure weight. This, together with the uplift forces, due to the pressure intensities on the base area of

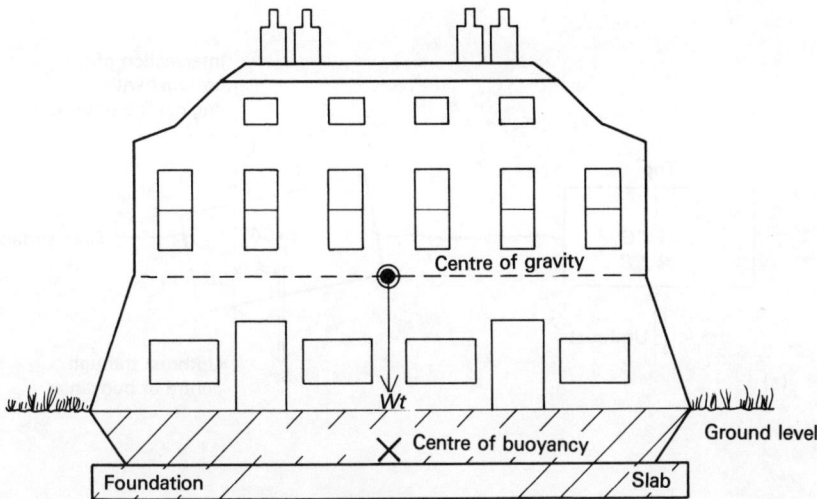

Fig. 4.2 Large building founded on saturated soil

the foundations, makes the structural design of buildings on such sites complex.

Where buoyancy effects exist, it is important to realise that the structure involved behaves much as a ship partially immersed in the sea.

To ensure that the structure is stable, it is important that it should be able to right itself, if for some reason it is displaced.

For stable equilibrium a self-righting couple is needed. In the case depicted (Fig. 4.3), this is provided by the weight (Wt) times the lever arm distance (L) to the line of action of the buoyancy upthrust.

In unstable equilibrium, no such self-righting couple exists and the body would simply topple. It can be seen (Fig. 4.3b) that the action of the weight on its lever arm (L) would indeed cause the body to continue to roll over.

In general, when the line of action of the upthrust, through the centre of buoyancy, cuts the original line of action of the weight, above the centre of gravity, a body is stable. Conversely, when the two lines of action intersect below the centre of gravity, the body is unstable. If the intersection coincides with the centre of gravity, then the body is in neutral equilibrium and the body will remain in any position to which it is displaced.

Example 4.2

What relationship exists between the submerged weight of a body, immersed in a fluid, and the difference in hydrostatic pressure forces on its upper and lower surfaces?

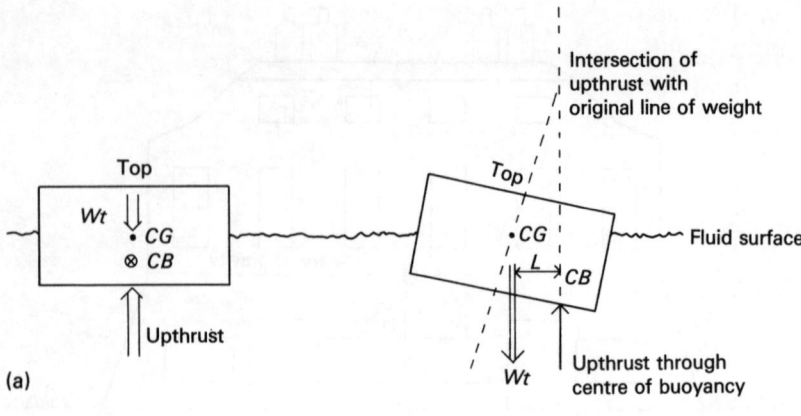

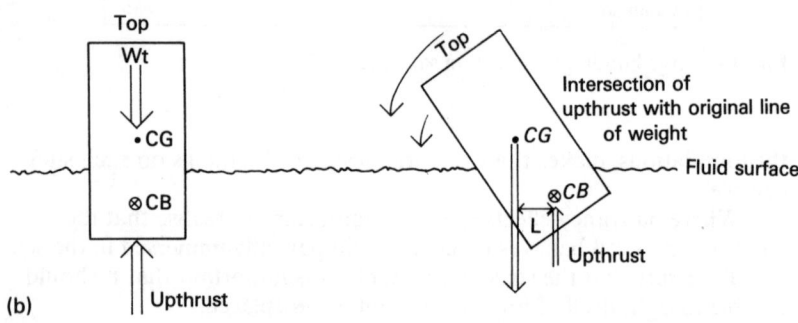

Fig. 4.3 Conditions for stability and unstability in floating bodies. (a) Stable equilibrium (self righting couple set up when body displaced) (b) Unstable equilibrium (overturning couple set up when body displaced)

Take as an example a rectangular steel tank (1 m high, 2 m wide and 3 m long) that weighs 98.1 kN in air, and which floats in fresh water with its upper surface 2 m below the water level.

Solution

From Archimede's Principle, the submerged weight of the tank is:
submerged weight = air weight − weight of volume of water displaced by tank,

i.e.

$$= 98.1 \text{ kN} - \frac{1000}{1000} \times 9.81 \times (1 \times 2 \times 3) \text{ kN}$$

$$= 98.1 \text{ kN} - 58.86 \text{ kN}$$

$$= 39.24 \text{ kN}$$

If the forces that act on the tank are considered, it is apparent that the only three that exist are –

- the air weight of the tank, acting downwards,
- the hydrostatic force of a 2 m depth of fresh water on the tank's upper surface, also acting downwards,
- the hydrostatic force of a 3 m depth of fresh water on the tank's lower surface, and this acts upwards.

Since the plan areas of the tank's upper and lower surfaces are both $2 \times 3 \text{ m}^2$, these forces are

air weight	–	98.1 kN ↓
hydrostatic force (upper surface)	–	117.72 kN ↓
hydrostatic force (bottom surface)	–	176.58 kN ↑

Thus a remnant force of 39.24 kN (acting downwards) exists.

This exactly equals the submerged weight of the tank. Had any other depths of submergence been taken an identical result would be found.

So the submerged weight of a body in a fluid exactly equals its air weight minus the difference in hydrostatic pressure forces on its lower and upper surfaces.

Example 4.3

A factory is being built on a riverside site, a section of which is shown on Fig. 4.4.

The foundation slab (constructed from concrete whose density is 2600 kg/m³) has a total plan area of 40 m by 40 m.

Groundwater in this area is at its lowest level (G.W.L.–Fig. 4.4) in the autumn and rises to ground surface in mid-winter. The water's density is 1050 kg/m³.

The foundation slab was completed in October, and, as the client wishes to delay the remainder of the construction until the following spring, the options are

(a) to leave the construction as drawn in Fig. 4.4
    or
(b) to build perimeter walls around the slab up to ground level to ensure that the surrounding soil does not slump over the slab, as the groundwater level rises.

Which suggestion is preferable?

Solution
Option (a). If the construction is left as drawn, then the rising groundwater will ultimately fill the excavation above the foundation slab. In

48

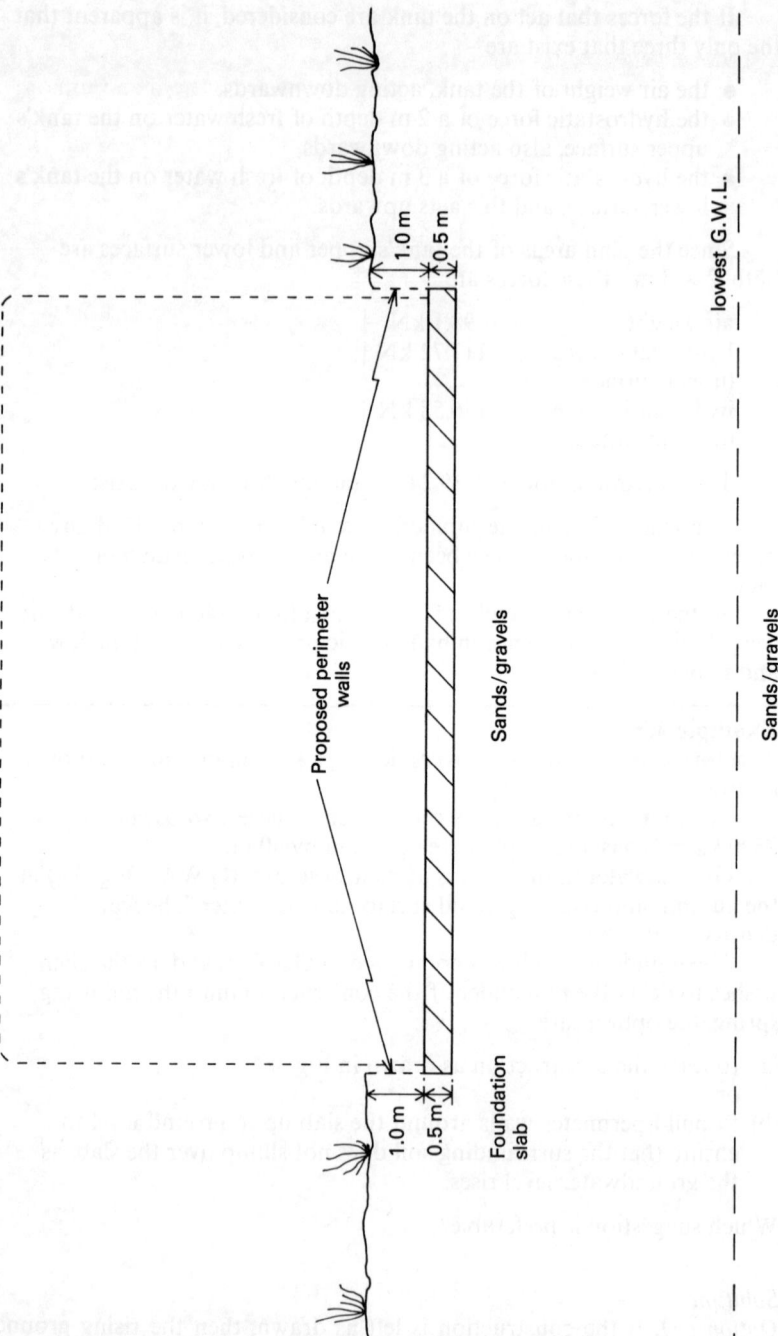

Fig. 4.4

this case, the forces which will enter an analysis are:–

- the weight of the slab
- the difference in hydrostatic pressure on the bottom and top of the slab.

The volume of the slab is $\quad 40 \times 40 \times 0.5 \text{ m}^3$

its mass is $\qquad\qquad\quad 40 \times 40 \times 0.5 \times 2600 \text{ kg}$

and so its weight is $\qquad 40 \times 40 \times 0.5 \times \dfrac{2600}{1000} \times 9.81 \text{ kN}$

$= \underline{20\,404.8 \text{ kN}} \quad \text{(acting downwards)}$

The difference in hydrostatic pressure on the bottom and top of the slab is

$dP = \rho \times g \times 1.5 \times \text{area} - \rho \times g \times 1.0 \times \text{area}$

$\quad = \rho \times g \times 0.5 \times \text{plan area of slab}$

$\quad = \dfrac{1050}{1000} \times 9.81 \times 0.5 \times 40 \times 40 \text{ kN}$

(assuming that the groundwater density is 1050 kg/m^3)

$= \underline{8240.4 \text{ kN}} \quad \text{(acting upwards)}$

Thus, in this case, a remnant force of

$20\,404.8 - 8240.4 \text{ kN}$

i.e. $\quad 12\,164.4 \text{ kN}$

still acts vertically downwards, and the foundation slab will not experience any disturbance from hydraulic factors.

Option (b). If perimeter walls are built up to ground level, two possible conditions can occur.

If the walls allow groundwater to leak through them, hydrostatic conditions will be exactly as those calculated in option (a), and the slab will remain unaffected. However, if the walls are impermeable barriers to the entry of groundwater, and if rainfall does not fill the excavation above the slab, the forces which have to be considered are – as before –

- The weight of the slab
- the difference in hydrostatic pressure forces on the bottom and top of the slab.

As, in the worst case, no water will lie above the slab, a hydrostatic pressure force, on the base of the slab, of

$P = \dfrac{1050}{1000} \times 9.81 \times 1.5 \times 40 \times 40 \text{ kN}$

$\quad = 24\,721.2 \text{ kN}$

will exist. This is in excess of the air weight of the slab and so the foundation would attempt to float upwards, straining and ultimately breaking any drains, water supply and other connections already built into the slab.

An examination of Fig. 4.4 reveals the reason for this situation. The foundation slab and the perimeter walls together act exactly as does a boat placed on water.

From a viewpoint of hydraulics, option (a) is thus preferable. If, however, the costs and inconvenience of clearing slumped soil from above the foundation slab, in spring, are so significant that perimeter walls have to be constructed, then they should be built with gaps left to allow the rising groundwater to fill the excavation above the slab.

Chapter 5

Fluids in motion—
the complication of
viscosity

5.1 The relative imprecision of hydrodynamics

All the examples, in the earlier chapters, yielded precise solutions by the application of a simple and limited amount of theory. If more complicated examples of, for example, the fluid forces on bent and twisted surfaces of variable cross-sectional area had been attempted, even these could have been exactly solved by the same methods. It would only have been additionally necessary to project these irregular geometric shapes on to chosen and mutually perpendicular axes.

This typifies hydrostatics, which is an exact science, simply because only a single type of force (due to fluid weight and density) is being considered.

In contrast, hydrodynamic problems can be analysed only to an approximate accuracy, and the examples given in the following chapters will, to some degree, be of situations simplified from reality. This will be most apparent in the treatment of open channel flow (Ch. 11) where the bias will be to man-made channels of regular geometric cross-sections. Natural river channels, whose cross-sectional areas and bed roughnesses vary significantly from place to place on the river, are simply too complex for the theory that is available.

The reason for this relative imprecision is that additional forces appear in fluid motion and these prevent an engineer having the same detailed knowledge as he has in hydrostatic problems. As a result, the theory available in hydrodynamics often has to be supplemented by experimental evidence. This point will be stressed through the following text.

Despite the difficulty of analysing detail, it should not be believed that major engineering works cannot be safely and precisely designed. The available theory applied with sound judgement and commonsense is adequate, and many practising engineers prefer hydrodynamic to hydrostatic design, simply because it does offer the opportunity for personalised judgement.

5.2 Viscosity

The reason, above all others, for the difficulty in obtaining precise details of a hydrodynamic problem is the existence of a property – **viscosity** – possessed by all fluids to some extent.

Viscosity is the ability of a fluid to stick to solid surfaces (the walls of a pipe, the bed of a channel, the edges of a bridge pier, etc.) and to exert a drag on them, which in turn has to be overcome by the using up of some of the energy in the fluid.

Very viscous liquids, such as tars, oils, glues, paints and treacle, are thick, slow-flowing fluids which cling to any solid surface. Anyone who has stirred a tin of paint will realise how difficult it is to make the paint run off the stirring rod.

Other fluids such as air and water appear not to show this effect, but even these low-viscosity fluids exert a drag on solid boundaries. If this were not so, then balloons and boats would remain stationary in a wind or in a moving river, without the need for anchors.

The practical effect of viscosity – which of course does not occur in hydrostatic problems where no relative movement between the fluid and its container takes place – is that the elements of the fluid closest to the solid boundaries are slowed down by viscous drag on the boundaries. This produces the situation that fluid particles at increasing distance from the boundaries move at greater and greater velocities. The zone of near stationary fluid against the solid surface is termed the **boundary layer** and can vary from a few millimetres to several metres in thickness, depending on the fluid's viscosity and the roughness of the solid surface.

5.3 Boundary layers

As boundary layers can seldom be seen in everyday life, their visualisation presents problems for many students.

It is therefore worth carrying out a simple experiment (thus of course confirming the earlier statement that, in hydrodynamics, experimentation is more often a necessity than an exception).

A box, say 1 m long, 0.200 m wide and 0.100 m deep, is constructed so that the end face can be raised to act as a gate (Fig. 5.1).

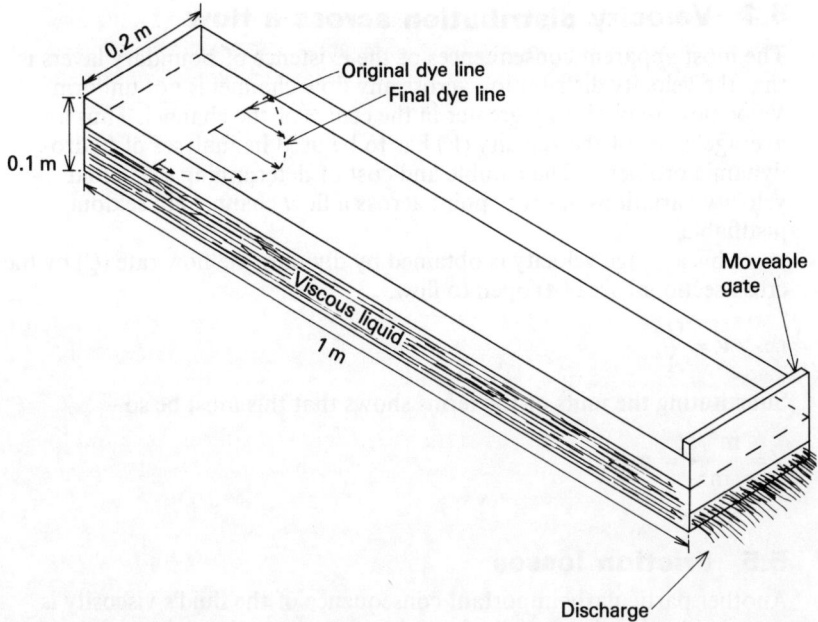

Fig. 5.1 Boundary layer experiment

The box is half filled with any available viscous liquid (water thickened with dissolved sugar is suitable, as is household emulsion paint thinned with water) and a dye line is drawn on the liquid's upper surface. When the end gate is lifted, some liquid runs out and the dye line moves forward along the box. If the gate is then closed, the initial and final dye line positions can be compared and it will be obvious that the liquid adjacent to the sides of the box has hardly moved. The thickness of the boundary layer can then be measured.

If the experiment, which takes only a few minutes to complete, is repeated with liquids of different viscosity (made simply by dissolving different weights of sugar in the same volume of tap water) the variation of boundary layer thickness with viscosity can be established and it will be found that boundary layers thicken as viscosity increases.

If more time is available, it is worth coating the inside of the box with a rough material (sandpaper can be used) and repeating the experiment with an already tested viscous liquid. In this case, the additional roughness of the solid boundaries offers greater opportunity for the fluid to stick to the walls and a thicker boundary layer will be produced.

More detailed consideration of boundary layers will be necessary when the effects of these on bridge piers is considered (Ch. 11).

5.4 Velocity distribution across a flow

The most apparent consequences of the existence of boundary layers is that the velocity distribution across any flow channel is not uniform. Velocities are obviously greater in the centre of the channel. Thus an average value of the velocity ($\bar{V}$) has to be used in analyses of hydrodynamic problems. The trouble and cost of determining the actual velocity variations at every point across a flow channel are seldom justifiable.

This average velocity is obtained by dividing the flow rate (Q) by the cross-sectional area (A) open to flow.

i.e. $\bar{V} = \dfrac{Q}{A}$

Substituting the units of the terms shows that this must be so –

$$\bar{V} = \frac{m^3/s}{m^2} = \underline{m/s}$$

5.5 Friction losses

Another particularly important consequence of the fluid's viscosity is that the slowing down of the boundary layers requires a loss of some part of the energy possessed by the fluid. This energy is obviously lost counteracting the viscous drag of the fluid on its confining walls and is generally termed the **friction loss**, or the **major loss**, or simply **friction**. It is an inevitable fact that some energy is always lost in fluid flow and the major portion of this loss is due to the viscous drag of the fluid.

Frictional losses are a major preoccupation of hydraulic engineers, either because the flow of the fluid has been powered with expensive pumped energy or because a client needs a particular fluid energy at the end of the pipe or channel.

It has already been shown, in the boundary layer experiment, that the roughnesses of the solid boundaries increase viscous drag and thus these rougher surfaces are said to have a higher frictional loss. However, almost all surfaces we use for pipes and channels can have their surface roughness varied with usage and time. Ordinary metals (steel and iron) rust and roughen in contact with air and water and it should not be surprising that unprotected metal walls of a pipe can give greatly increased frictional losses after a few years of use. In other cases, the fluid passing an initially rough surface can coat and smooth it, leading to a lower frictional loss with time. The walls of concrete sewers often show this effect.

The problem for a designing engineer working with a real fluid, which has viscosity, is to predict the frictional loss the fluid will create on channel walls of a particular material and the change, if any, of this frictional factor with time. A great deal of experimental research has gone to produce the tables of frictional loss.

5.6 Laminar and turbulent states of flow

Another complicating effect of viscosity is that it alters the way fluids actually flow. This property was discovered by Osborne Reynolds, a pioneer hydraulic researcher, in 1884. Reynolds carried out a simple experiment with small bore, smooth glass pipes draining a tank, where a constant depth of water was maintained by a controlled inflow (Fig. 5.2)

The flow path of the water was traced out by dye from a syringe.

Reynolds noted that at low velocities of flow (of the order of 30 mm/s) the dye marked out a clear, sharp line. As the flow velocity was increased to around 75 mm/s, by simply increasing the constant depth of

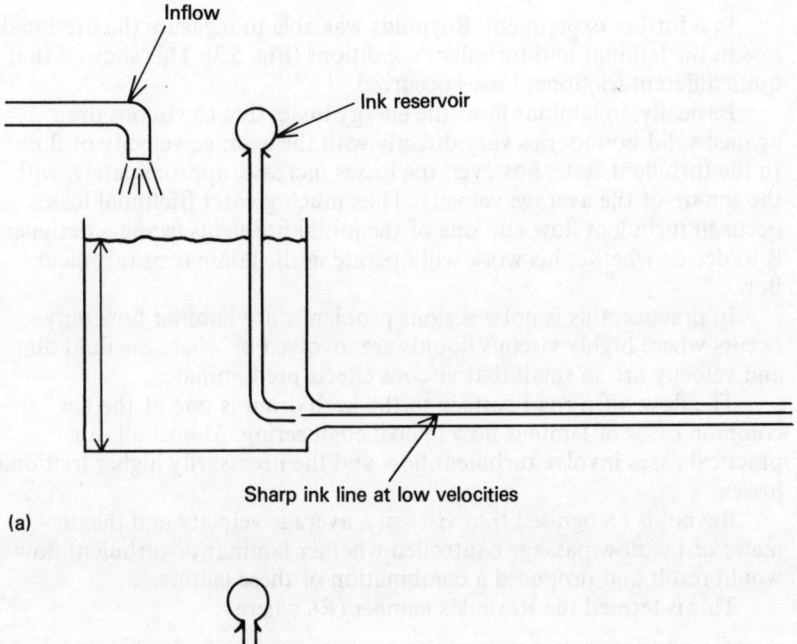

(a)

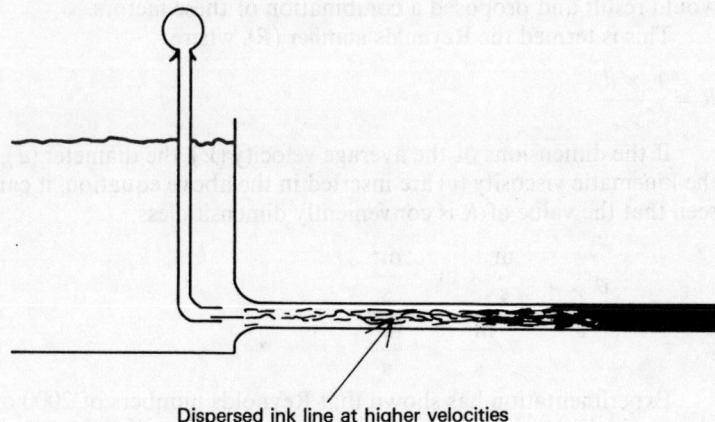

(b)

Fig. 5.2 Osborne Reynolds' experiments

water in the tank, the dye streak became hazy and dispersed through the water column.

If the average flow velocity were reduced back to the initial value, once again a clear, sharp dye line resulted.

The only possible explanation for this is that fluids can flow in two quite different states –

(a) as **laminar flow** where the elements of water slide smoothly past each other, rather like plates of glass
and
(b) as **turbulent flow** where transverse eddies and swirls develop, mixing the elements of the flow across the channel.

In a further experiment, Reynolds was able to measure the frictional loss in the laminar and turbulent conditions (Fig. 5.3). This showed that quite different frictional losses occurred.

Basically, in laminar flow, the energy losses due to viscous drag against solid boundaries vary directly with the average velocity of flow. In the turbulent state, however, the losses increase, approximately, with the square of the average velocity. Thus much greater frictional losses occur in turbulent flow and one of the initial problems facing a designer is to decide whether his work will operate in the laminar or turbulent flow.

In practice, this is not a serious problem since laminar flow only occurs where highly viscous liquids are involved or where the fluid film and velocity are so small that viscous effects predominate.

The flow off a road surface to the kerb drain is one of the few common cases of laminar flow in civil engineering. Almost all other practical cases involve turbulent flow and the necessarily higher frictional losses.

Reynolds recognised that viscosity, average velocity and the diameter of the flow passage controlled whether laminar or turbulent flow would result and proposed a combination of these factors.

This is termed the **Reynolds number** (R), where,

$$R = \frac{\bar{V} \times d}{v} \qquad\qquad\qquad [5.1]$$

If the dimensions of the average velocity ($\bar{V}$), the diameter (d), and the kinematic viscosity (v) are inserted in the above equation, it can be seen that the value of R is conveniently dimensionless,

i.e. $R = \dfrac{\bar{V} \times d}{v} = \dfrac{\dfrac{m}{s} \times m}{\dfrac{m^2}{s}} = \dfrac{\dfrac{m^2}{s}}{\dfrac{m^2}{s}}$

Experimentation has shown that Reynolds numbers of 2000 or less always signify laminar flow, whilst values of 4000 or above typify the turbulent state.

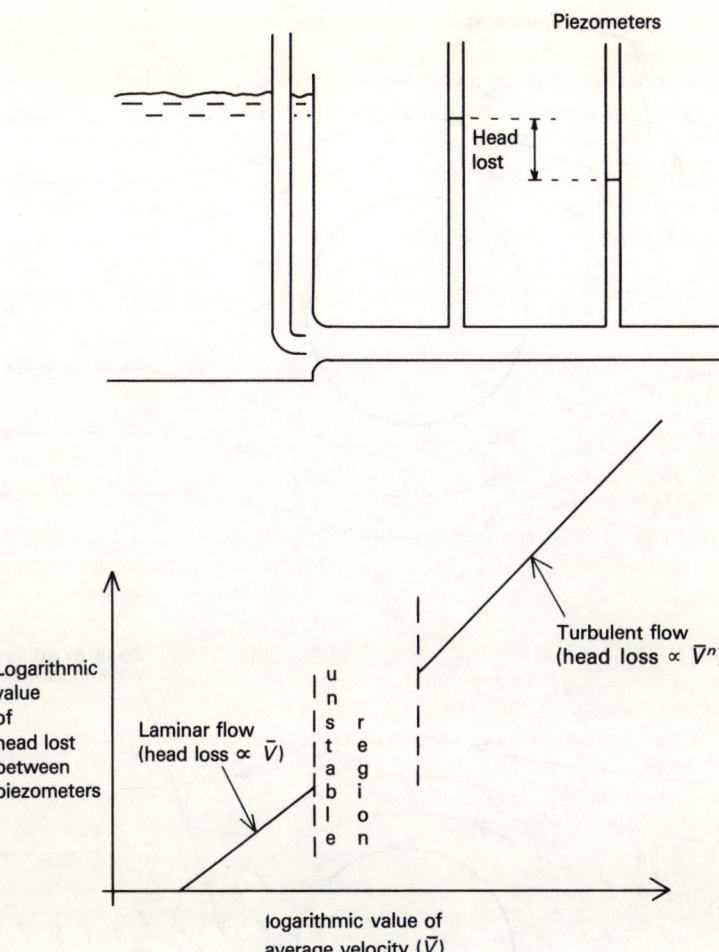

Fig. 5.3 Head lost in laminar and turbulent flow

5.7 Turbulence and minor losses

A further complicating effect of viscosity can be seen when the behaviour
of a flowing fluid around a solid body (e.g. the pier of a bridge or a
partly closed valve in a pipe) is examined (Fig. 5.4).

If the fluid is ideal (i.e. lacking any viscosity) the pattern of the
distortion of the streamlines is identical both upstream and downstream
of the solid body. (Streamlines are the paths marked out by arbitrarily
chosen particles of the moving fluid and can be seen only by placing
some visual tracer – dyes or small bright particles of metal – in the fluid.)

58

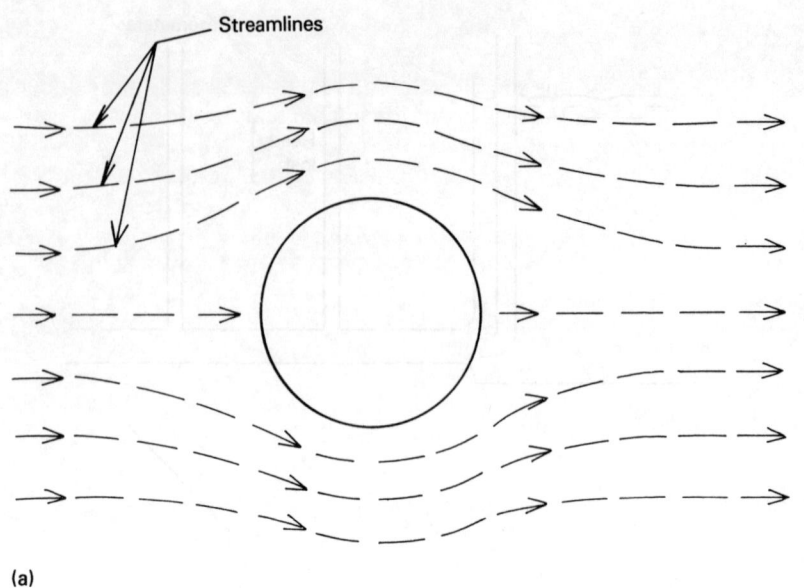

Streamlines

(a)

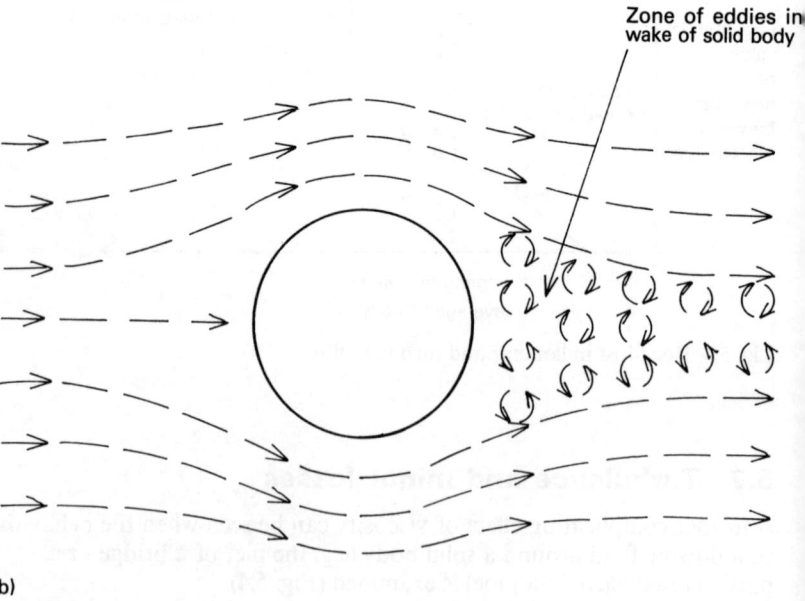

Zone of eddies in wake of solid body

(b)

Fig. 5.4 Difference between the flow of ideal and real fluids past a solid body. (a) Frictionless fluid (b) Real viscous fluid (additional energy losses due to eddies)

With a real fluid, however, the stream line distortion downstream of the solid body is quite different.

The reason for this simply is that all real fluids possess quite high moments of inertia and as a result are unable to describe tight curves. This leaves a zone of dead fluid in the wake of the solid body. The viscous drag of the moving fluid on this dead zone produces turbulent eddies, whose rotation is powered by energy taken from the moving fluid.

Such turbulent eddies are the cause of the **minor losses** which always occur as localised energy losses at any point where the stream lines of the moving fluid are abruptly distorted. In some design situations these minor losses can be important and it is generally found desirable to disrupt the smooth flow of a fluid as little as is possible to avoid this additional loss of energy. More will be made of this when the design of pipelines is considered (Ch. 7).

5.8 Summary of viscosity effects

The existence of fluid viscosity is the reason above all others for the lack of precise knowledge of how a fluid will actually behave in particular flow channels. The pattern of velocity variation across the channel, the amount of frictional loss due to viscous drag on the channel walls, the state of flow that will result and the localised losses of energy due to the production of turbulent eddies, all result from the fluid having a particular viscosity.

Other viscosity effects, such as the drag forces on solid bodies, have not been discussed and will be treated at a later stage.

Despite this major complicating factor, methods do exist to allow an adequate, if somewhat generalised, knowledge of fluid movement to be determined.

Self assessment questions

1. Is hydrodynamics less precisely understood than hydrostatics because it is less important and because less investigation has been carried out?
2. Define the fluid property of 'viscosity' and give the units in which it is expressed.
3. List the complications created by the existence of viscosity in all fluids.
4. Will a boundary layer be thicker or thinner if the fluid is made more viscous?
5. The smoothness of the solid boundaries exposed to the flow of a fluid does not affect the thickness of the boundary layer produced. Is this true?
6. Is frictional loss a property of the fluid or of the solid surface over which it flows?

7. Can friction change with time? Give examples.
8. Explain clearly what is meant by the word 'turbulent' when it is applied to fluid flow. Is the fluid in turbulent flow disturbed by waves, eddies and cross currents?
9. Would you expect the drainage off a flat roof to be turbulent?
10. Where do minor losses occur?

Examples

Example 5.1

What diameter of pipe is needed to pass 6.500 m³/s at an average velocity of 3 m/s? If the water is at a temperature of 50 °C (kinematic viscosity at this temperature is 0.556×10^{-6} m²/s) is the flow laminar or turbulent?

Solution

(a) Since $Q = \bar{V} \times A$ or $A = \dfrac{Q}{\bar{V}}$

$\therefore$ Area needed for the flow $= \dfrac{6.500}{3} = 2.167$ m²

and since the area of a circular section pipe is

$\dfrac{\pi \times D^2}{4}$

$\therefore$ the pipe diameter wanted is

$D = \left(\dfrac{2.167 \times 4}{\pi}\right)^{1/2} = 1.661$ m

(b) The flow state is found by applying [5.1], that is

$R = \dfrac{\bar{V} \times d}{v} = \dfrac{3 \times 1.661}{0.556 \times 10^{-6}} = 8.962 \times 10^6$

This value is far in excess of the critical 4000, above which turbulent flow must occur.

In fact, the specified average velocity of flow should have indicated that turbulent flow would occur, since laminar flow is only possible at low velocities.

Example 5.2

If laminar flow had been required in Q. 5.1, to minimise the energy losses in friction, what average flow velocity would have to occur? Assume the same pipe diameter and water temperature.

Solution

For laminar flow, the Reynolds number must not exceed a value of 2000.

$$\therefore \quad R = 2000 = \frac{\bar{V} \times d}{v}$$

or $\quad \bar{V} = \dfrac{2000 \times 0.556 \times 10^{-6}}{1.661} = 669.476 \times 10^{-6} \text{ m/s}$

$$= 0.000669 \text{ m/s}$$

or approximately half a millimetre per second.

This velocity is so slow as to be practically insignificant. Almost every practical flow situation requires higher velocities and so operates in the turbulent state.

Chapter 6

Methods of analysing fluid flows

Despite the complexities created by viscosity, a surprisingly large proportion of practical problems are solvable by three simple techniques – the continuity, Bernoulli and momentum equations.

6.1 The continuity equation

The **continuity equation** applies only to those cases where the flow rate (Q) does not change with time. Such flows are termed steady and whilst these are less common than unsteady flows, i.e. where the flow varies, a consideration of steady flows offers the easiest introduction to hydrodynamics.

In steady flows, as Q is invariant, the inflow to any channel must equal the outflow from it, thus

$Q_{inflow} = Q_{outflow}$

and since

$Q = \bar{V}A$

$\therefore \quad (\bar{V}A)_{inflow} = (\bar{V}A)_{outflow}$

or in general

$$Q = \bar{V}_1 A_1 = \bar{V}_2 A_2 = = = \bar{V}_n A_n \qquad [6.1]$$

at any chosen n points along the channel.

The value of this simple technique for practical application can scarcely be overestimated, allowing, as it does, the calculation of average flow velocities at any chosen point in a network of channels, provided that the flow rate and the cross-sectional areas of the flow channels are known.

Section 6.1 – Self assessment questions

1. Clearly distinguish between 'steady' and 'unsteady' flows.
2. Give at least one case of unsteady flow with which you are familiar.
3. Can the continuity equation be applied to any flow?

Example 6.1

A pipeline, made of 500 mm internal diameter (I.D) pipe, which transmits a flow of 0.050 m³/s, has to be bridged over a stream. To reduce the load on the bridge, and thus its costs, it is decided that 200 mm internal diameter pipe will be used in the bridged length of the pipeline.

What difference in average velocity of flow will this create?

Solution

Using 200 mm I.D. pipe

$$Q = 0.050 \text{ m}^3/\text{s} = V \times A$$

$$= V\left(\pi \times \frac{D^2}{4}\right)$$

$$= V\left(\pi \times \frac{0.2^2}{4}\right)$$

or the average velocity of flow is 1.591 m/s

Had 500 mm I.D. pipe been used on the bridge

Q would still be 0.050 m³/s

$$\therefore \quad 0.050 = V \times A$$

$$= V \times \left(\pi \times \frac{0.5^2}{4}\right)$$

from which, the average velocity of flow, using 500 mm I.D. pipe, would have been 0.255 m/s.

The difference in the velocity over the bridged section will be 1.591 − 0.255 m/s

$$= 1.336 \text{ m/s}$$

Such variations in pipeline diameter are often necessary to reduce the load on weak ground or on structures. The only limitation on their use is whether or not the increased average velocity of flow will create additional hydraulic problems (see section 6.3 on momentum forces).

Example 6.2

A domestic hosepipe is passing 400 litres/hour of water. The hose has a 12 mm internal diameter and the nozzle at the end of the hose has a 3 mm diameter jet. What is the average velocity in the hose pipe and the jet?

Solution

$Q = 400$ litres/hour $= 0.111 \times 10^{-3}$ m^3/s

(since 1000 litres $= 1$ m^3)

as the flow is steady

$Q = (V \times A)_{hose} = (V \times A)_{nozzle}$

Velocity in hosepipe

$$V_{hose} = \frac{Q}{A_{hose}} = \frac{0.111 \times 10^{-3}}{\dfrac{\pi \times 0.012 \times 0.012}{4}}$$

$$= \frac{0.111 \times 10^{-3}}{\dfrac{3.168 \times 10^{-3}}{28}}$$

$$= 0.981 \text{ m/s}$$

Velocity in nozzle

$$V_{nozzle} = \frac{Q}{A} = \frac{0.111 \times 10^{-3}}{\dfrac{\pi \times 0.003 \times 0.003}{4}} = 15.7 \text{ m/s}$$

Example 6.3

The earliest Suez Canal was built about the year 1500 BC as a water-way some 64 km long and with a cross-sectional area of 300 m^2. The Mediterranean Sea has a higher water level than the Red Sea, thus a slight flow from North to South took place along the canal.

The canal silted up and became disused, largely because windblown desert sand settled in the channel.

Could the canal have been designed to avoid siltation?

Assume (a) that desert sand particles

(i) settle in water whose velocity of flow is 0.010 m/s or less;

(ii) are carried along by water flowing at 0.200 m/s or more.

(b) that ships of the period had drafts of up to 2 m and widths of up to 5 m.

Solution

Since the canal did silt up, we may assume that the average velocity of flow was, at most, 0.010 m/s.

To avoid siltation, any sand particles blown into the canal should be carried along with the flow. Thus the cross-sectional area of flow should have been reduced, so:

$(V \times A)$ original $= (V \times A)$ redesigned

or

$0.010 \times 300 = 0.200 \times A$ (redesigned)

from which the redesigned area $= 15 \text{ m}^2$

Given the dimensions of the ships the canal had to accommodate, this redesign would only have been successful if the canal operators had been willing to give up moving ships in both directions at the same time.

A valid point comes out of this example: a design is only successful if the engineer's client finds it satisfies his working requirements.

6.2 The Bernoulli equation

The **Bernoulli equation** is essentially a re-statement of the general principle of conservation of energy in fluid flow.

That energy (or the capacity to do work) is conserved and can be changed from one form to another is familiar from everyday life, where examples, such as the release of stored solar energy (in, say, coal or oil) to boil water and raise steam, which then can propel a ship, are commonplace.

Daniel Bernoulli's real contribution was to determine that fluid energy can appear in three distinct forms – as kinetic (or velocity), pressure, and elevation (or positional) energy.

These three forms all have the dimensions (units) of length and are often termed the kinetic, pressure and elevation heads. Although all are familiar phenomena in everyday life, it is worth considering a little further what each actually is, since much of a hydraulic engineer's work consists of altering one type of energy to another to produce some desired practical result.

If a mass of fluid (M) moves at some velocity (V) it can be shown that its kinetic energy (K.E.) per unit weight of fluid is:

$$\frac{M \cdot \dfrac{V^2}{2}}{M \cdot g}$$

where $M \cdot g$ is of course the product of the fluid's mass and the gravitational attraction (i.e. the weight of the fluid),

or $\quad$ K.E. $= \dfrac{V^2}{2g}$

Substituting the units for these terms shows that

$$\text{K.E.} = \frac{m^2}{s^2} \bigg/ \frac{m}{s^2}$$

$\quad\quad = m \quad$ (i.e. the unit of length)

Similarly the pressure energy at some chosen point in a fluid where the pressure is p and the cross-sectional area of flow is A has the units of length, as is outlined below.

Work done per second by the fluid = the pressure force exerted by the fluid on the cross-sectional area of flow, multiplied by the distance a cross-sectional flow element moves in 1 second

$$= (p \cdot A)(V)$$
$$= (p \cdot A)(Q/A)$$
$$= p \cdot Q$$

But the weight of fluid passing this cross-sectional element in a second $=$ the mass of the fluid times the gravitational attraction

$$= M \cdot g$$
$$= (\rho \cdot Q)g$$

Thus the pressure energy per unit weight of fluid $= \dfrac{pQ}{\rho Q g} = \dfrac{p}{\rho g}$

which as has already been shown (Ch. 2) to have the dimensions of length.

Finally, the potential energy of any mass of fluid held at a height Z (metres) above a datum level is

its weight times the elevation distance

or

$(M \cdot g)(Z)$

and the potential energy (P.E.) per unit weight of fluid is

$\therefore \quad$ P.E. $= \dfrac{Mgz}{Mg} = Z$ (metres)

The inter-relations between these forms of fluid energy can be seen if a tank (full of a fluid under some constant pressure) is pierced to allow a water spout to rise (Fig. 6.1).

At point A, immediately before the fluid emerges from the hole in the tank, the fluid has the same elevation as the chosen datum (Z^1-Z^1) and thus no elevation energy exists; its flow velocity is still negligible (thus the kinetic energy is zero) and all the available energy exists as pressure head.

At point B, the fluid has just emerged as a jet into the atmosphere and thus has to have the same pressure as the surrounding environment (therefore its gauge pressure is zero and its pressure energy relative to the atmosphere is also zero). As the elevation of B is essentially that of the datum, the elevation energy is also zero and all the available energy exists as velocity head.

Finally at point C, where the jet has reached its maximum height and the fluid has slowed to a halt, the kinetic energy has returned to a zero value and the pressure energy (relative to the atmosphere) is zero, as it was at B. Therefore all the available energy exists as the positional form.

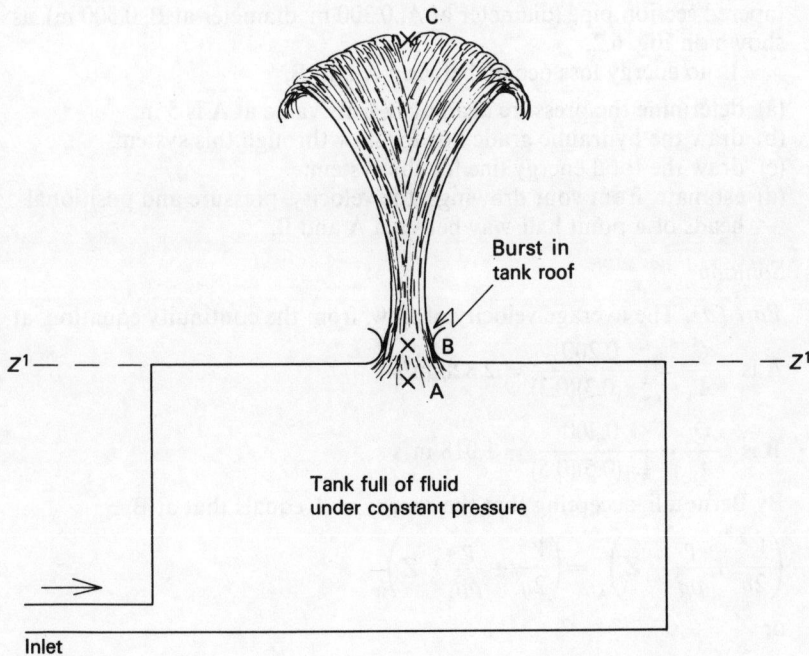

Fig. 6.1 Forms of fluid energy

If no energy loss takes place, due to viscous drag and the resultant heating up of the interface between the fluid and the atmosphere, then

the energy at A = the energy at B = the energy at C

or

the pressure head = the velocity head = the elevation head

In practice, this result is unusual, since it is more normal for a fluid, at each of the points of interest, to possess more than one type of energy and, of course, some energy is always lost in fluid flow.

In general, therefore, the Bernoulli equation is expressed as:

$$\left(\frac{V^2}{2g} + \frac{p}{\rho g} + Z\right) \text{ at A} = \left(\frac{V^2}{2g} + \frac{p}{\rho g} + Z\right)$$

at B + the energy losses between A and B   [6.2]

The Bernoulli equation is of greater application than the Continuity equation and is the most useful analytical tool for those hydrodynamic problems where the energy loss between the points of interest, in a fluid system, is gradual and relatively small. Many pipe, open channel and pump problems fall into this category.

Example 6.4
Water flows from A to B at a rate of 0.200 m³/s in an 8 m-long tapered section pipe (diameter at A, 0.300 m; diameter at B, 0.500 m), as shown on Fig. 6.2.

If no energy loss occurs between A and B,

(a) determine the pressure head at B, if its value at A is 5 m;
(b) draw the hydraulic grade line for flow through this system;
(c) draw the total energy line for the system;
(d) estimate, from your drawings, the velocity, pressure and positional heads of a point half way between A and B.

Solution

Part (a). The average velocity of flow, from the continuity equation, at

A is $\quad \dfrac{Q}{A_A} = \dfrac{0.200}{\frac{1}{4}\pi(0.3)(0.3)} = 2.828$ m/s

B is $\quad \dfrac{Q}{A_B} = \dfrac{0.200}{\frac{1}{4}\pi(0.5)(0.5)} = 1.018$ m/s

By Bernoulli, accepting that the energy at A equals that at B,

$$\left(\frac{V^2}{2g} + \frac{p}{\rho g} + Z\right)_A = \left(\frac{V^2}{2g} + \frac{p}{\rho g} + Z\right)_B$$

or

$$\left(\frac{2.828^2}{2g} + 5 + 3\right) = \left(\frac{1.018^2}{2g} + \frac{p}{\rho g} + 2\right)$$

(note that the centre-line of the pipe at A lies 3 m above the arbitrarily chosen datum, whilst the centre-line at B is 2 m above the datum. The choice of any other horizontal datum for the potential energy term would not affect the answer).

$$(0.408 + 5 + 3) = \left(0.053 + \frac{p}{\rho g} + 2\right)$$

or

$$\frac{p}{\rho g} \text{ at } B = 6.355 \text{ metres of water}$$

Part (b). The hydraulic grade line is simply the line tracing the variation in the sum of the elevation and pressure heads along the length of the pipe, and is drawn above the pipe for those cases where the pressure is greater than atmospheric, and below the pipe where sub-atmospheric, pressures (i.e. negative gauge pressures) occur. The shape of the hydraulic grade line need not fall continuously along the direction of fluid flow and can alter where the pressure conditions in the pipe are varied.

In this case, the hydraulic grade line can be taken as joining the values of pressure heads at A and B, as shown on Fig. 6.2.

Part (c). The total energy line simply traces out the variation in the sum of the elevation, pressure and velocity heads along the pipe and since, in this case, no energy loss occurs the total energy line is horizontal and lies a distance of 8.408 m above the chosen datum.

In practical cases involving real fluids, some energy would of course be lost in viscous drag and turbulent eddies and the total energy line would fall continuously in the direction of flow. An increase in the total energy at any downstream point in the flow system is only possible if additional energy is added to the fluid at that point (as would occur for example if a centrifugal pump were installed at that point in the pipe).

Part (d). The required values of the kinetic, pressure and positional heads at a point halfway between A and B are, from Fig. 6.2,

elevation head = 2.500 m
pressure head = 5.700 m
velocity head = 0.208 m

The value of the hydraulic grade and total energy lines are apparent from this example. If the lines can be drawn from information at any two points in the fluid system, then the energy conditions at any other point on the system can be read directly off the drawing.

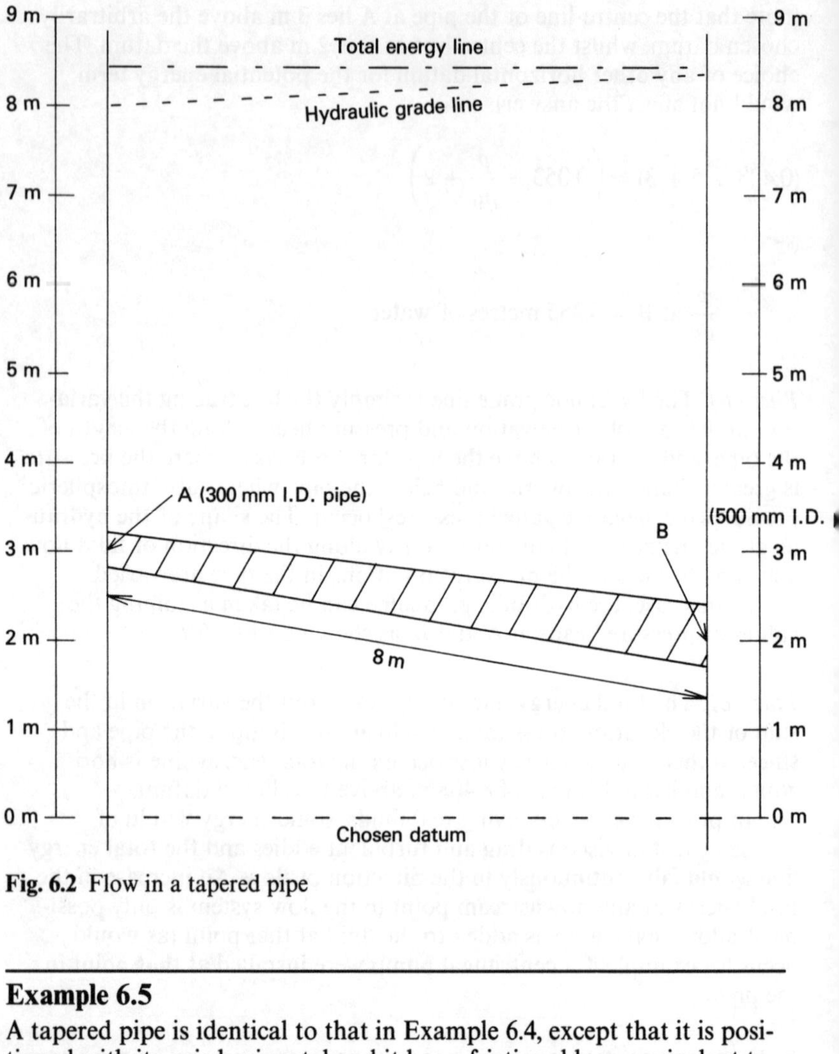

Fig. 6.2 Flow in a tapered pipe

Example 6.5

A tapered pipe is identical to that in Example 6.4, except that it is positioned with its axis horizontal and it has a frictional loss, equivalent to 0.11 m of water for each metre of pipe length.

(a) determine the pressure head at B in metres of water;
(b) draw the hydraulic grade and total energy lines

Solution

(a) As this flow is more realistically accompanied by a loss of energy, the Bernoulli equation can be written

$$\left(\frac{V^2}{2g} + \frac{P}{\rho g} + Z\right)_A = \left(\frac{V^2}{2g} + \frac{P}{\rho g} + Z\right)_B + \text{energy losses from A to B,}$$

i.e. accepting the previously calculated velocities

$$\left(\frac{2.828^2}{2g} + 5 + 0\right) = \left(\frac{1.018^2}{2g} + \frac{P}{\rho g} + 0\right) + \text{losses}$$

since the Z terms are identical for a horizontal pipe

or $(0.408 + 5) = \left(0.053 + \frac{P}{\rho g}\right) + 0.88$ m

since an 8 m length of pipe will have a head loss equivalent to
0.88 m of water,

∴   the pressure head at B = 4.475 m of water.

(b) The hydraulic grade line is drawn taking the centre line of the pipe
as the datum. The pressure head at A is 5 m and that at B is
4.475 m. Thus the hydraulic grade line falls from A to B.
The total energy line lies above the hydraulic grade line, at a distance equivalent to the velocity head. Thus its value at A is 5.408 m
and at B is 4.528 m.
The fall of the total energy line in the direction of flow is typical of
real-life flows.

Example 6.6
A reservoir is compelled by law to pass a constant supply of 0.150 m³/s
to the river below the dam wall.
   To do this a 0.200 m internal diameter horizontal pipe, 243 m long,
is laid (Fig. 6.3). The frictional losses with this pipe are expected to be
0.62 m of water for each 10 m length of pipe.
   What head of water must always be held in the reservoir to ensure
that this outflow is achieved?

Solution
Applying the Bernoulli equation between points A and B we obtain,

$$\left(\frac{V^2}{2g} + \frac{P}{\rho g} + Z\right)_A = \left(\frac{V^2}{2g} + \frac{P}{\rho g} + Z\right)_B + \text{losses between A and B}$$

But, as the pipe is horizontal, the Z terms cancel out,

or:

$$\left(0 + \frac{P}{\rho g}\right)_A = \left(\frac{V^2}{2g} + 0\right)_B + 15.066 \text{ m}$$

Since the water at A is hardly moving, its velocity head is negligible. Similarly, at B, the water discharges to the atmosphere before
reaching the river below and the pressure head at B is therefore gauge
zero.

72

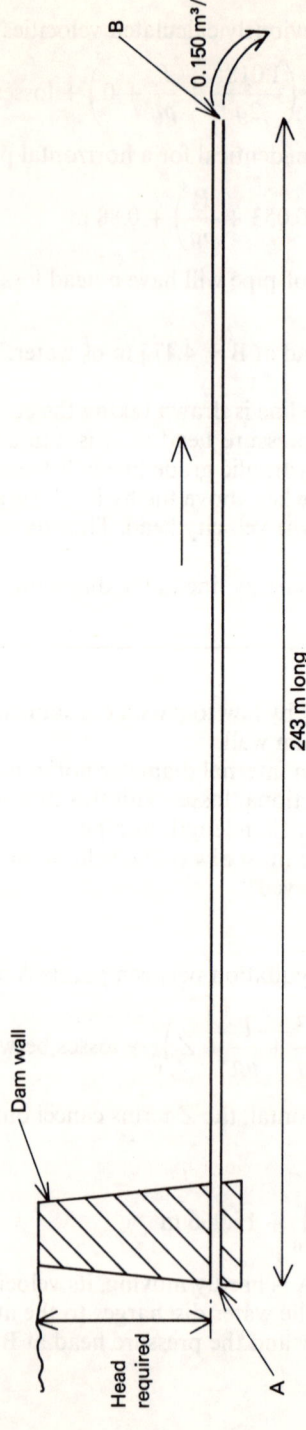

Fig. 6.3 Discharge from a reservoir

B

0.150 m³/s

243 m long

Dam wall

Head required

A

With a flow rate of $0.150 \text{ m}^3/\text{s}$ and a pipe of 0.200 m internal diameter, the average velocity at B is

$$\frac{Q}{A} = \frac{Q}{\pi D^2/4} = 4.773 \text{ m/s}$$

and the velocity head $= \dfrac{V^2}{2g} = 1.161$ m of water

$\therefore$ the pressure head at A $= 16.227$ m,

as the pressure head $= \dfrac{P}{\rho g} = h,$

where h is the depth of water creating the hydrostatic pressure, the depth of water in the reservoir must never fall below 16.227 m above the centre of the drain pipe.

Example 6.7

The discharge from a small amenity dam is via a concrete circular section syphon, whose diameter is 0.25 m (Fig. 6.4). The concrete is relatively rough and gives a frictional head loss of 1.10 m between A and B and 3.85 m between B and C.

   Determine:

(a) the maximum discharge through the syphon if the reservoir and river levels can be assumed as constant;
(b) the fluid pressure intensity in the crest of the syphon.

Solution

(a) Applying Bernoulli between points A and C we obtain

$$\left(\frac{V^2}{2g} + \frac{P}{\rho g} + Z\right)_A = \left(\frac{V^2}{2g} + \frac{P}{\rho g} + Z\right)_C + \text{losses (A} - \text{C)}$$

At A, the water is under a pressure head equivalent to 2 m of water, and a potential energy of 8 m, above the datum of the dam base. The water just about to enter the syphon has a negligible velocity and so its velocity head may be ignored. Thus, at A, the total energy $= (0 + 2 + 8) = 10$ m of water.

   At C, the water is under a pressure of 1 m of water and lies 1 m above the chosen datum for potential energy. Additionally, the water just about the emerge from the syphon mouth has a velocity head of

$$\frac{V^2}{2g}$$

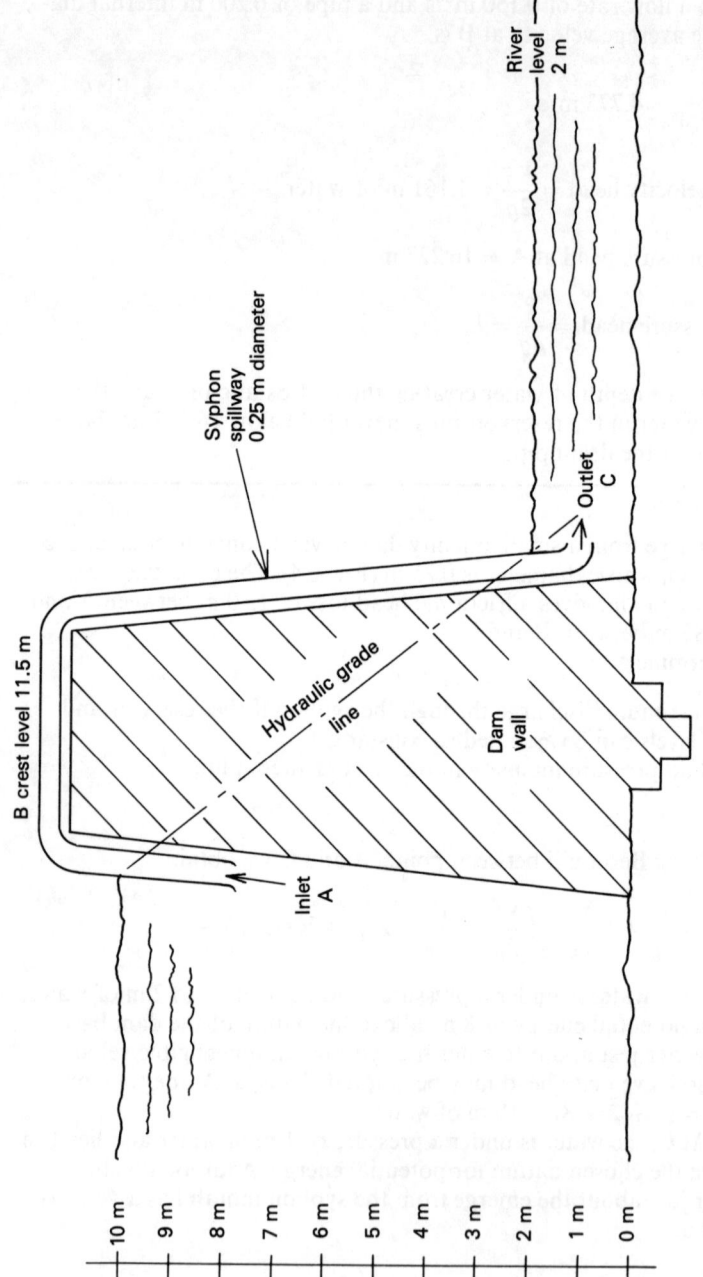

Syphon spillway 0.25 m diameter

B crest level 11.5 m

Hydraulic grade line

Dam wall

Inlet A

Outlet C

River level 2 m

10 m
9 m
8 m
7 m
6 m
5 m
4 m
3 m
2 m
1 m
0 m

Fig. 6.4

Thus at C, the total energy $= \left(\dfrac{V^2}{2g} + 1 + 1\right)$,

and since the losses due to frictional viscous drag on the syphon walls between A and C are equivalent to 4.95 m of water, the energy balance between A and C is

$$10 \text{ m} = \left(\dfrac{V^2}{2g} + 2\right) + (4.95)$$

or

$$\dfrac{V^2}{2g} = 3.05 \text{ m of water}$$

$$\therefore \quad V = (2g \cdot (3.05))^{1/2} = 7.736 \text{ m/s}$$

$$\therefore \quad \text{the discharge } Q = \text{V.A.} = (7.736)\left(\pi \times \dfrac{0.25^2}{4}\right)$$

$$= 0.380 \text{ m}^3/\text{s}$$

(b) to obtain the pressure head at B, the Bernoulli equation is applied between points A and B.
The total energy at B is

$$\left(\dfrac{V^2}{2g} + \dfrac{P}{\rho g} + Z\right) = \left(3.05 + \dfrac{P}{\rho g} + 11.5\right)$$

and the energy loss due to friction between A and B is 1.10 m,

$$\therefore \quad \text{equating the energy at the two points gives}$$

$$10 \text{ m} = \left(\dfrac{P}{\rho g} + 14.55\right) + 1.10$$

or the pressure head at the crest of the syphon

$$= -5.65 \text{ m of water.}$$

Pressures at the crests of syphons are necessarily lower than atmospheric to ensure that flow (which only takes place from a high pressure head to a lower energy state) will actually occur. The hydraulic grade line would join the water levels in the reservoir and the river (see Fig. 6.4) and a large part of the syphon will be seen to lie above the grade line and thus has sub-atmospheric pressures. A limit on how far a syphon can lie above the hydraulic grade line is imposed by a property of water called cavitation (see Ch. 10). As the pressure of water falls, initially any dissolved air is released (usually at about -7.5 m of pressure head at sea-level). At a lower pressure equivalent to a negative atmosphere (-10.3 m of water at sea-level) the water liquid abruptly vapourises to a gas. Both effects lead to gas blockages in the pipeline.

The other practical problem of installing a syphon is to 'prime' it. Priming consists of expelling all air from the syphon and filling it with the fluid to be piped. This usually is achieved by valves at the inlet and outlet to the syphon.

Section 6.2 – Self assessment questions

1. List the three forms in which fluid energy can occur and show for each that the energy is measured in metres length of the fluid concerned.
2. Define clearly what is meant by the term 'hydraulic grade line'.
3. If the top of a pipe, carrying a fluid, were pierced at four points to allow the fluid to escape, how far would the fluid jets rise?
4. Is the hydraulic grade line always drawn above the centre line of the pipe?
5. Where does the total energy line always lie?
6. With real fluids (i.e. fluids which possess viscosity) does the total energy line always slope down in the direction of flow?
7. Explain clearly why a syphon is able to cause a discharge of a fluid.

6.3 The momentum equation

The Bernoulli equation is only applicable where the energy losses between the points of interest can be quantified.

When abrupt localised energy losses occur or where there is a particularly complicated flow condition, which precludes details of the energy loss being found, some other analytical technique is needed.

The **momentum equation** satisfies this need. This equation is directly derived from Newton's Second Law of Motion, which can be written as:

The force applied to moving a body in a certain direction is equal to the rate of change of the body's momentum in that direction

Momentum is defined as the product of the body's mass and its velocity,

i.e.   Momentum $= M \times V$

and in fluid units $= \dfrac{\text{mass}}{\text{volume}} \times \dfrac{\text{volume of flow}}{\text{time interval}} \times \text{time interval} \times \text{velocity}$

$$= \rho \times Q \times dt \times V$$

or, substituting $V \times A$ for Q

$$= \rho \times A \times V^2 \times dt$$

Thus the rate of change of momentum is

$$\rho \times A \times V^2$$

and this equals the fluid force exerted in the direction of the momentum change.

If the case of flow through a tapered pipe is considered (Fig. 6.5a), it is obvious that the average velocity of flow varies from one end of the pipe to the other and so a momentum change takes place. Thus a force exists because of this and is

$$F = \rho(A_2 V_2^2 - A_1 V_1^2) \qquad\qquad [6.3]$$

Where a flow channel not only tapers but also bends (Fig. 6.5b) it is necessary to calculate the force in both the X- and Y-planes by taking the

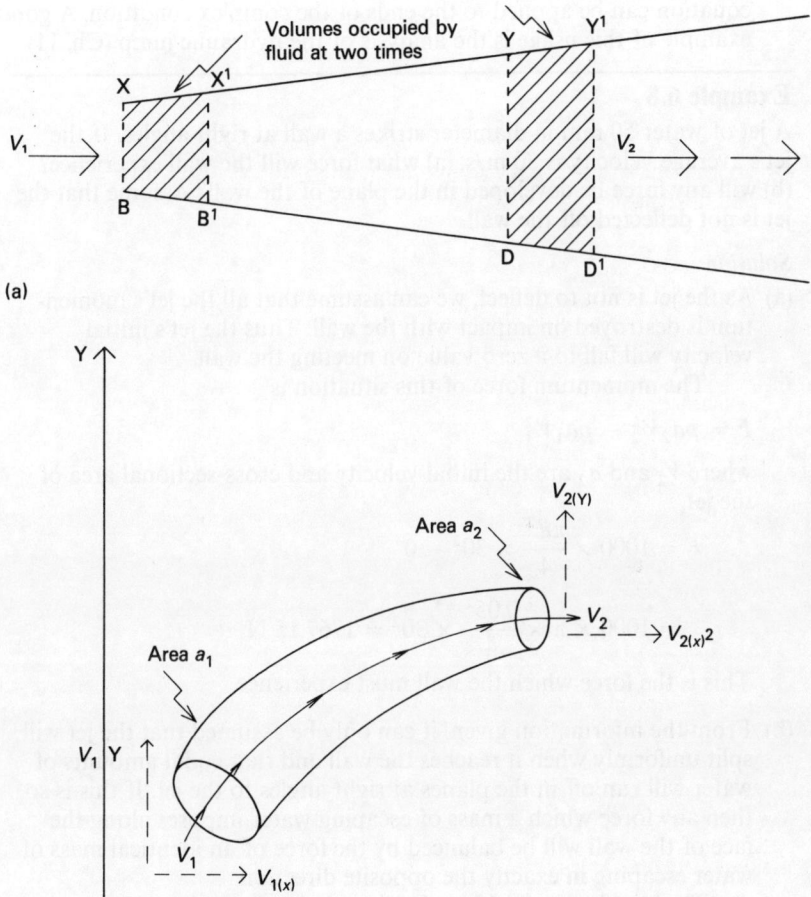

(a)

(b)

Fig. 6.5 (a) Flow in a tapered pipe with velocity variation (b) Flow in a tapered and bent pipe

velocity components in these directions,

i.e. $F(\text{X-plane}) = \rho(a_2 V_2^2 - a_1 V_1^2)\text{x-plane}$

and $F(\text{Y-plane}) = \rho(a_2 V_2^2 - a_1 V_1^2)\text{y-plane}$

The two common applications of the momentum equation are:

(a) where the forces a fluid exerts on its boundaries are wanted. These are simply the reactions to the calculated momentum forces and have the same numerical values;

(b) where a complex flow condition exists, within which no precise knowledge of energy changes are available. Here the momentum equation can be applied to the ends of the complex condition. A good example of this usage is the analysis of the hydraulic jump (Ch. 11).

Example 6.8

A jet of water 50 mm in diameter strikes a wall at right angles. If the jet's average velocity is 30 m/s, (a) what force will the wall experience? (b) will any force be developed in the plane of the wall? Assume that the jet is not deflected off the wall.

Solution

(a) As the jet is not to deflect, we can assume that all the jet's momentum is destroyed on impact with the wall. Thus the jet's initial velocity will fall to a zero value on meeting the wall.

The momentum force of this situation is

$$F = \rho a_2 V_2^2 - \rho a_1 V_1^2$$

where V_2 and a_2 are the initial velocity and cross-sectional area of the jet,

$$\therefore \quad F = 1000 \times \frac{\pi d^2}{4} \times 30^2 - 0$$

$$= 1000 \times \pi \times \frac{0.05^2}{4} \times 30^2 = 1767.15 \text{ N}$$

This is the force which the wall must experience.

(b) From the information given, it can only be assumed that the jet will split uniformly when it reaches the wall and that equal amounts of water will run off in the planes at right angles to the jet. If this is so, then any force which a mass of escaping water imposes along the face of the wall will be balanced by the force of an identical mass of water escaping in exactly the opposite direction.

Thus no forces would be developed along the plane of the wall.

This example is a good illustration of the need for experimental investigations in hydrodynamics. Until we carry out such an experiment and measure how much of the water runs off in each direction on the plane of the wall, we cannot factually attempt to answer the problem.

Example 6.9

Had the jet, in Example 6.8, struck the wall at an angle other than 90°,

(a) develop equations for the forces normal and parallel to the wall;
(b) identify the additional data you would require to answer the question.

Solution

(a) As the jet does not strike the wall exactly at right angles, it cannot be assumed that it will still be equally diverted along the wall.

From Fig. 6.6 it is obvious that more of the fluid will flow along the right-hand part of the wall.

This example includes not only a variation in the fluid's velocity but also a change in the fluid's direction of flow. It is therefore necessary to calculate the momentum forces in two mutually perpendicular planes. The plane of the wall, and a plane at right angles to it, are convenient.

At right angles to the wall, the component of the incoming jet's velocity is

$V \cos \alpha$

Thus the momentum force (remembering that all the momentum is destroyed at contact with the wall) normal to the wall is

$F = \rho(a_2 V_2^2 \cos \alpha - 0)$

In the plane of the wall itself, the momentum force *appears* to be

$\rho \cdot a \cdot V^2 \cdot \sin \alpha$

as $V \sin \alpha$ is the component of the jet's velocity parallel to the wall.

However, the wall's rough surface and the viscosity of the fluid will ensure that shear forces will occur. These inevitably will reduce the fluid's velocity on the wall.

(b) To solve this problem it will be necessary to set up the experiment shown in Fig. 6.6. From this, the average velocities of flow along the wall could be obtained and the forces parallel to the wall quantified.

Example 6.10

A horizontal pipe is necessarily bent through 90°. What forces will this create and how can the stability of the pipe be ensured?

Solution

The pipe bend is shown on Fig. 6.7.

The fluid can be assumed to have some particular average velocity – $\bar{V}$ – and pressure intensity, p.

Impact of jet at angle 90°

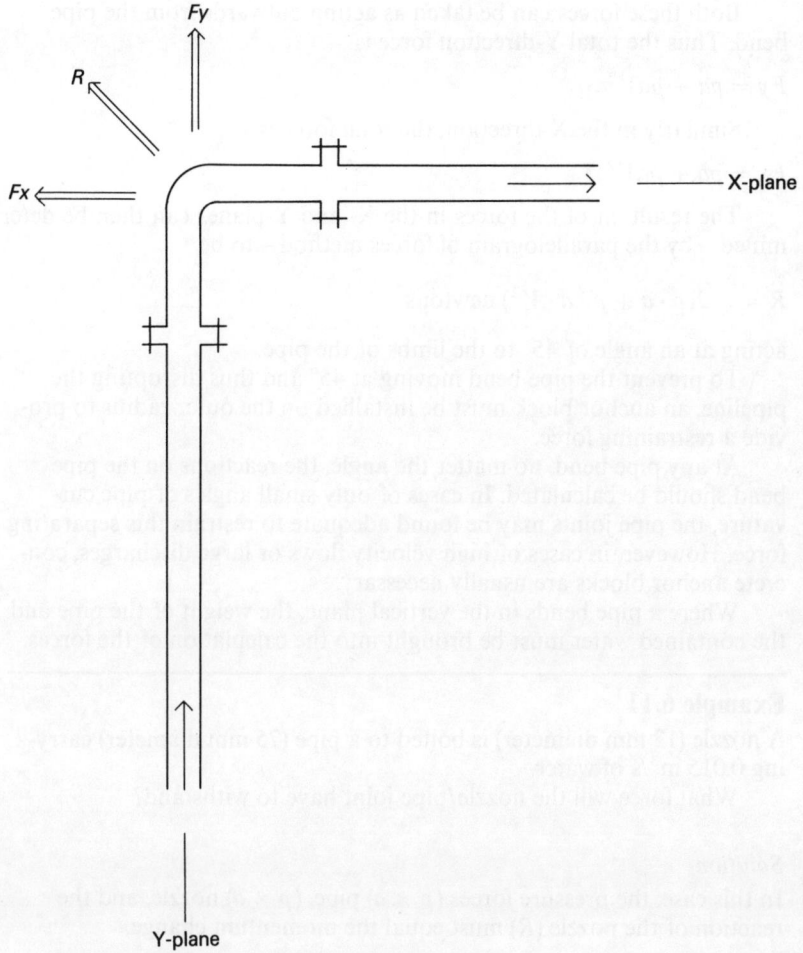

Fig. 6.7 Forces due to momentum change in a 90° bend

In view of the change in direction of fluid flow, it is necessary to consider the various fluid forces with respect to mutually perpendicular axes. The obvious axes to choose are those of the pipe limbs.

In the Y-direction, at the pipe bend, two forces occur–due to hydrostatic and momentum factors.

The hydrostatic force (see Ch. 2) is simply

$F_1 = p \cdot a$, where a is the pipe's cross-sectional area.

The momentum force

$F_2 = \rho \cdot a \cdot V^2$,

since all velocity in the Y-direction vanishes at the pipe bend.

Both these forces can be taken as acting outwards from the pipe bend. Thus the total Y-direction force is

$$Fy = pa + \rho aV^2$$

Similarly in the X-direction, the total force is

$$Fx = pa + \rho aV^2$$

The resultant of the forces in the X- and Y-planes can then be determined – by the parallelogram of forces method – to be

$$R = \sqrt{2}(p \cdot a + \rho \cdot a \cdot V^2) \text{ newtons}$$

acting at an angle of 45° to the limbs of the pipe.

To prevent the pipe bend moving at 45° and thus disrupting the pipeline, an anchor block must be installled on the outer radius to provide a restraining force.

At any pipe bend, no matter the angle, the reactions on the pipe bend should be calculated. In cases of only small angles of pipe curvature, the pipe joints may be found adequate to restrain this separating force. However, in cases of high velocity flows or large discharges, concrete anchor blocks are usually necessary.

Where a pipe bends in the vertical plane, the weight of the pipe and the contained water must be brought into the calculation of the forces.

Example 6.11

A nozzle (12 mm diameter) is bolted to a pipe (75 mm diameter) carrying 0.015 m³/s of water.

What force will the nozzle/pipe joint have to withstand?

Solution

In this case, the pressure forces ($p \times a$) pipe, ($p \times a$) nozzle, and the reaction of the nozzle (R) must equal the momentum change,

i.e. $\rho(aV^2 \text{ nozzle} - aV^2 \text{ pipe}) = (p \times a)\text{pipe} - (p \times a)\text{nozzle} - R$

or $\rho(aV^2 \text{ nozzle} - aV^2 \text{ pipe}) = (p \times a)\text{pipe} - R$

since the pressure intensity at the exit from the nozzle is atmospheric and so has a gauge value of zero.

From the continuity equation,

$$Q = (V \times A)_{\text{pipe}} = (V \times A)_{\text{nozzle}}$$

$$\therefore \quad 0.015 = V_{\text{pipe}} \times \pi \times \frac{0.075 \times 0.075}{4}$$

or $V_{\text{pipe}} = \underline{3.395 \text{ m/s}}$

similarly $V_{\text{nozzle}} = \underline{132.629 \text{ m/s}}$

The momentum force $= \rho(aV_{nozzle}^2 - aV_{pipe}^2)$

$\qquad\qquad\qquad = \rho Q(V_{nozzle} - V_{pipe})$

$\qquad\qquad\qquad$ since $Q = VA_{nozzle} = VA_{pipe}$

$\qquad\qquad\qquad = 1000 \times 0.015\,(132.629 - 3.395)$

$\qquad\qquad\qquad = 1.9385$ kN

The pressure intensity in the pipe can be obtained by applying the Bernoulli equation between the pipe and the nozzle – If energy losses can be ignored,

$$\frac{P_{pipe}}{\rho g} + \frac{V_{pipe}^2}{2g} = \frac{V_{nozzle}^2}{2g}$$

(since the pipe and nozzle are horizontal, the Z term is the same on both sides of the equation and can be ignored),

$$\therefore \quad \frac{P_{pipe}}{\rho g} = \left(\frac{132.629^2 - 3.395^2}{2 \times g}\right)$$

$$= 8\,789.5 \text{ kN/m}^2$$

and since the cross-section area of the pipe is

0.0044 m^2

then the pressure force in the pipe is

38.883 kN

Thus a tension force of

36.89 kN

is exerted by the nozzle on the nozzle/pipe joint.

Section 6.3 – Self assessment questions

1. Explain what 'momentum' means in terms of fluid parameters.
2. Distinguish clearly between situations where the momentum equation is more appropriate than the Bernoulli equation.
3. What do you understand by the term 'pipe bend reaction'? Why is an 'anchor block' often needed at pipe bends?

6.4 Summary of the analytical methods

The continuity equation relates velocities and cross-sectional areas of flow, the Bernoulli equation deals with the various types of fluid energy in these situations where energy losses are either relatively small and gradual or are quantifiable, and the momentum equation, relating as it does the changes in momentum to the forces producing these, allows the analysis of situations where more complex flow conditions occur.

These three methods permit the analysis of a large majority of hydraulic engineering cases.

Chapter 7

Gravity flow in pipes

7.1 The value of pipes

As a means of conveying fluids, pipes were developed historically later than open channels, simply because of the difficulty of firstly obtaining and then working the quantities of metal needed. However, once the necessary technology had been established, pipes became the preferred method in most cases. An obvious reason for this is seen where a fluid has to be carried over a hill (Fig. 7.1).

The costs and delays caused by cutting and filling the topography to the required base level of an open channel are high, unless a large and inexpensive labour force is available. In contrast, a pipe can be laid in a shallow trench over the hill and the required flow will take place, provided that the siphon can be primed (see section 6.2).

Pipelines can be powered by gravity, when the water level at the inlet is high enough above that at the outlet to provide the head required to overcome resistance to the flow. Where inadequate head exists naturally, pumped energy has to be supplied and such pipelines are termed 'pumped mains'. Gravity pipelines are not only historically the older but have a simpler and quite different method of analysis and design.

7.2 Basics of pipe flow

Within a pipeline, three zones are recognised – the pipe entry, the mid-pipe length, and the pipe exit. The types and magnitudes of energy losses differ in these three zones.

85

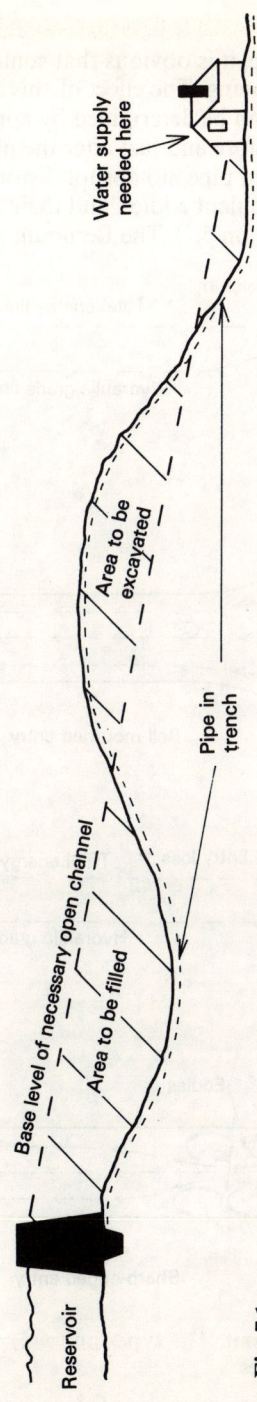

Fig. 7.1

Water supply needed here

Area to be excavated

Pipe in trench

Base level of necessary open channel

Area to be filled

Reservoir

7.2.1 Pipe entry

When a fluid enters a pipe, it is obvious that some acceleration has taken place to start the fluid moving. The effect of this acceleration on the energy level of the fluid can be determined by applying the Bernoulli equation to points just before and just after the pipe entry (Fig. 7.2).

With the bell-mouthed pipe, no abrupt distortion of the fluid's flow lines occurs. Thus no turbulent eddies, and their associated energy losses, take place (see section 5.7). The Bernoulli analysis is thus

$$\text{Energy}_{(\text{point A})} = \text{Energy}_{(\text{point B})}$$

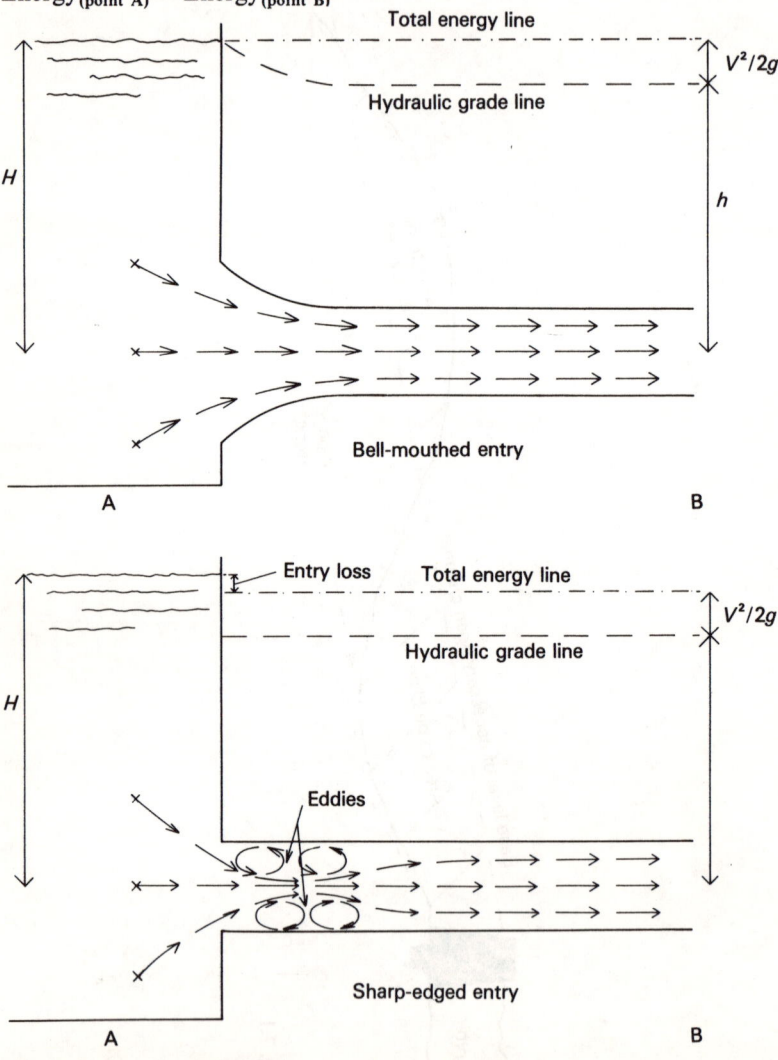

Fig. 7.2 Losses at pipe entries

(assuming that the length A–B is so short that significant frictional viscous losses do not occur),

or $\quad H = \dfrac{V^2}{2g} + h \quad$ where $\quad V =$ the average fluid velocity at **B**

and $\qquad\qquad\qquad\qquad h =$ the pressure head at **B**

$\therefore \quad \dfrac{V^2}{2g} = H - h$

Thus a diminution of the initial pressure head has had to take place to produce the average flow velocity. This is termed the entry loss.

In contrast, with a sharp-edged entry, abrupt curvature of the flow lines is inescapable. Thus the fluid stream is narrowed and flanked with turbulent eddies, which cause enhanced energy losses (section 5.7). The narrowing of the fluid stream is generally termed the **vena contracta**. A Bernoulli analysis, in this case, gives

$$H = \frac{V^2}{2g} + h + \text{turbulent eddy losses}$$

and so the entry loss to a sharp-edged pipe is always much greater (usually about ten times) than that to a bell-mouthed pipe.

Entry losses, caused by the need to accelerate still fluid into a pipe, are quite distinct from frictional effects and should not be confused with them.

7.2.2 The mid-pipe section

The earlier introduction to boundary layers (Ch. 5, Fig. 5.1) may not have made it obvious that they increase in thickness with distance of flow (Fig. 7.3a, b).

Whilst this growth is of great interest to those engineers who design aeroplanes and high-speed vehicles, of more obvious importance to the hydraulic engineer is what happens when a flat plate is wrapped around to make the normal circular-section pipe.

With boundary layers growing off the pipe walls, it is apparent that after a short distance of flow these layers will coalesce and completely fill the pipe's bore.

The length of this flow has been proved experimentally to be normally between thirty and fifty times the diameter of the pipe.

Once they join, the boundary layers completely fill the pipe and the velocity distribution remains constant at all downstream cross-sections. As the viscous drag of the fluid on the pipe walls has to remain at the value reached when the boundary layers coalesced, it can be said that

'for a particular flow rate, in a pipe of some fixed cross-sectional area and surface roughness, a constant frictional loss will be found for each unit length of the pipe. Thus a uniform hydraulic gradient will apply over the entire mid-pipe zone.'

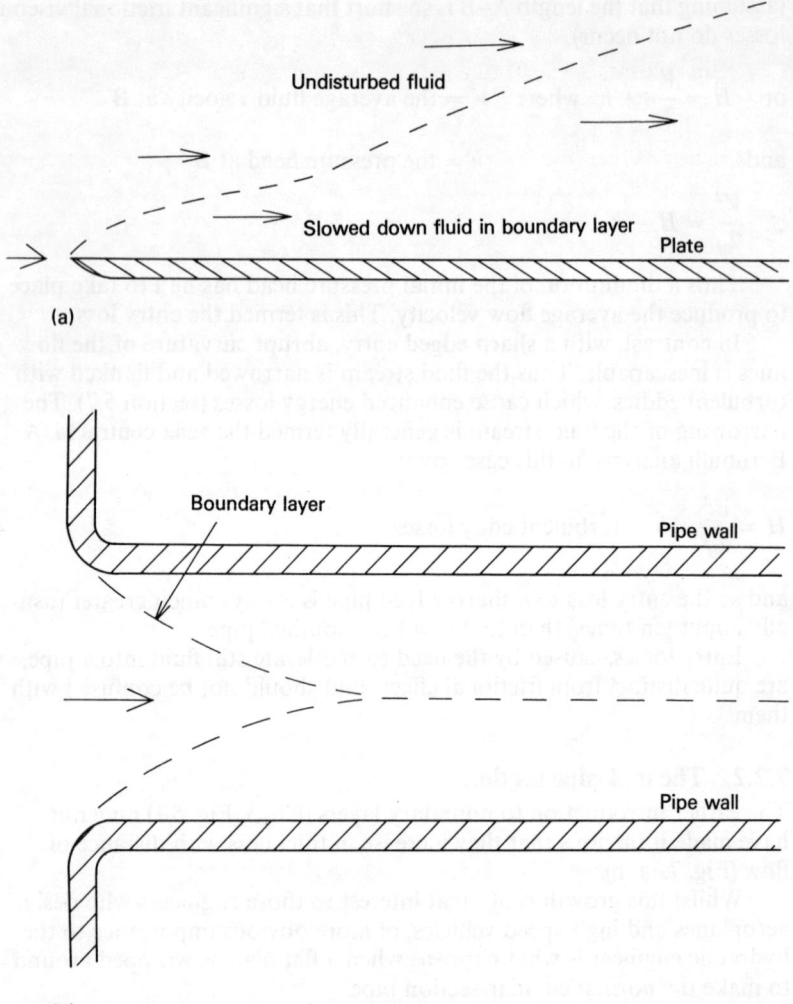

Undisturbed fluid

Slowed down fluid in boundary layer

Plate

(a)

Boundary layer

Pipe wall

Pipe wall

(b)

Fig. 7.3 (a) Boundary layer growth on flat plate (b) Boundary layer coalescing in circular pipe

This uniform hydraulic gradient will persist unless the pipe's cross-sectional area or roughness is altered. The viscous drag losses are the dominant, mid-pipe energy loss, and are termed the **major losses**.

It is worth stressing at this point, that the only type of energy the fluid has on entering a pipe is pressure head (see Fig. 7.2, point A). It is only this pressure head – after a portion of it has been subtracted to generate the full bore pipe velocity – that is available for degradation by frictional viscous drag. Thus the hydraulic gradient, once established, will remain constant throughout a pipe of uniform diameter and roughness. 'Major losses' are thus losses from the fluid's pressure energy.

'Minor losses' arise from a totally different cause. As the name suggests, they are usually of smaller importance than pipe friction.

Invariably such minor losses are localised reductions of the fluid's total energy, where the cross-sectional area open to fluid flow is abruptly altered in some way. The effect of this is to disturb the fluid's stream lines and to create the turbulent eddies noted in section 5.7 (Fig. 7.4).

Such turbulent eddy losses are localised to the particular points where the flow lines are disturbed. As a result, in a long pipeline, with relatively few valves, bends and changes in cross-section, they add up to an insignificant value compared to the total frictional losses. Conversely, a short pipeline will have a small overall frictional loss and, in this case, the minor losses may well be significant.

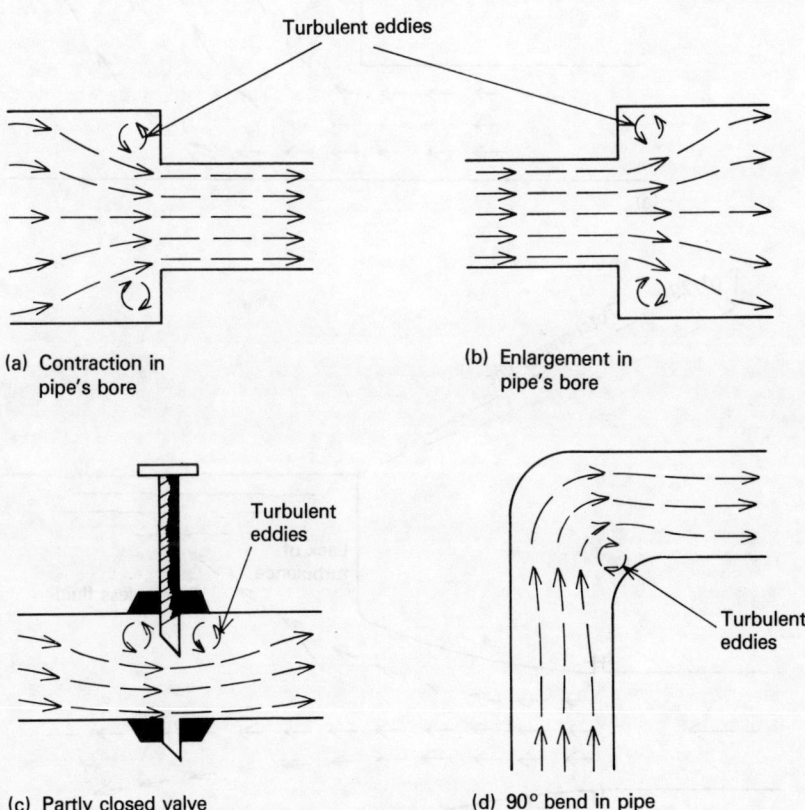

Fig. 7.4 Minor losses in a pipe flow. (a) Contraction in pipe's bore (b) Enlargement in pipe's bore (c) Partly close valve (d) 90° bend in pipe

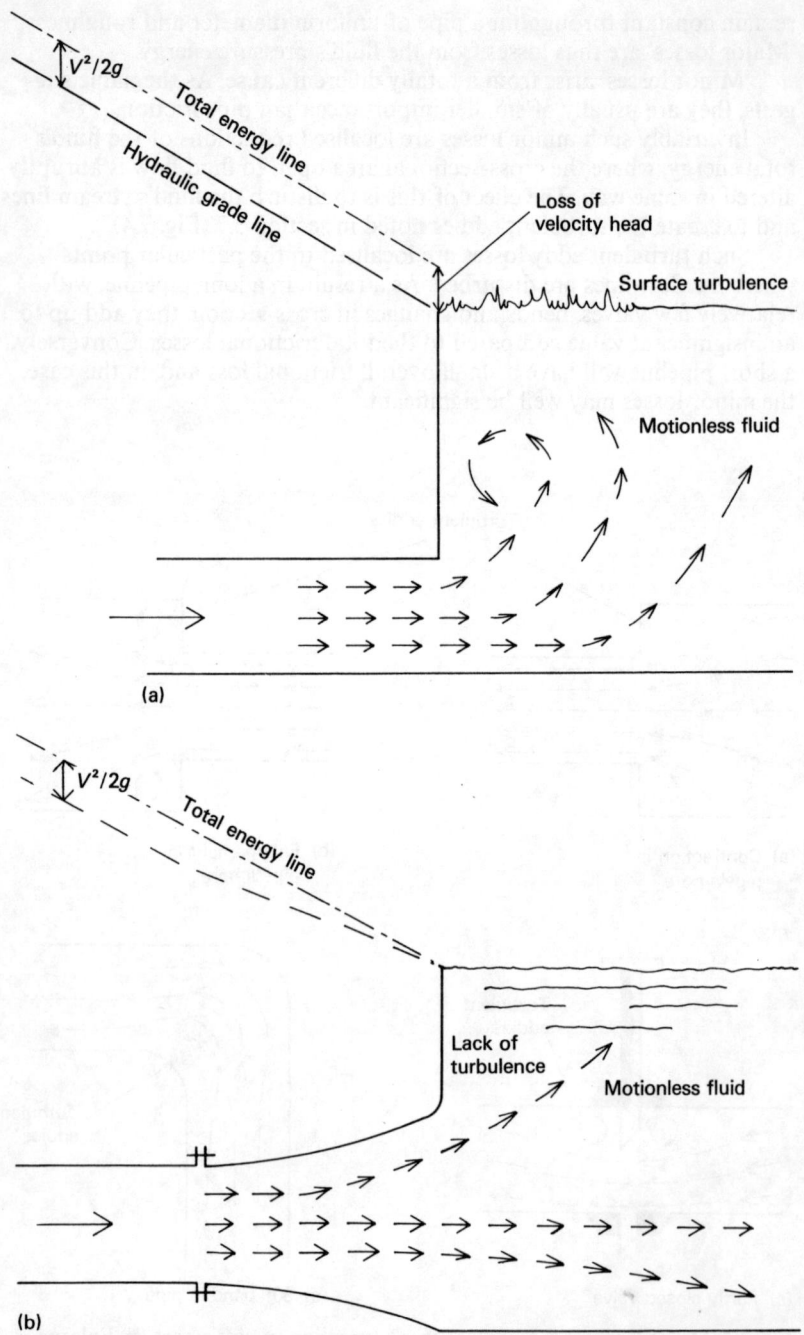

Fig. 7.5 (a) Sharpened edge pipe exit to tank (b) Tapered pipe exit to tank

7.2.3 Pipe exit conditions

If a pipeline has to discharge to the atmosphere, all the fluid's pressure head is necessarily lost.

Often, however, a pipeline terminates in a storage tank. In these instances, a relatively high speed fluid jet strikes a larger mass of still fluid. The consequences of this is that the fluid's velocity head is rapidly and completely destroyed, producing powerful eddies and heating up the fluid in the tank.

The loss of the entire velocity head can be that of a meaningfully large amount of head (for example, a fluid flowing at 6.260 m/s will have a velocity head of almost 2 m of water) and it is often worth while attempting to save this energy, to cut down the total energy requirement of a pipeline. The best method of achieving this aim is to gradually reduce the velocity – and thus the velocity head – of the fluid by passing it through a tapered pipe of increasing cross-sectional area, just before the storage tank (Fig. 7.5a, b).

In general, this method of gradually varying changes in the cross-sectional area open to the flow and thus distorting stream lines as little as possible, can be used to minimise most types of minor losses. Tapered sections joining alterations in pipe diameter and gradual bends in the pipe do reduce the minor losses and bring about savings in energy, though at the expense of greater capital outlay.

7.2.4 Minor losses in pipe flow

Over the greater part of a pipe's length, the flow area is full of boundary layer and a uniform value of frictional drag applies. In the relatively short entry and exit zones, losses of quite different origin occur.

Table 7.1 Minor losses in pipe flow

Fitting	$K \left(\text{where head loss} = K \cdot \dfrac{V^2}{2g} \right)$
Sharp edged pipe entry	– 0.5
Belled mouth entry	– 0.05
Gate valve (open)	– 0.20
Gate valve (half open)	– 5.60
Globe valve (open)	– 10.00
Sudden contraction of pipe diameter	– 0.44 × downstream velocity head
Sudden enlargement of pipe diameter	– $\dfrac{(V_1 - V_2)^2}{2g}$
Gradual enlargement of pipe diameter	– 0.02 to 0.04
30° pipe bend	– 0.07
45° pipe bend	– 0.25
90° pipe bend	– 1.15
Tee pieces	– 1.80

Turbulent eddy losses occur whenever fluid flow lines are disturbed and experimentally derived tables of such losses are available (Table 7.1).

Section 7.2 – Self assessment questions

1. Why does some pressure head always disappear at the entry to a pipe?
2. Explain, in terms of stream line distortion, why a bell-mouthed pipe entry creates a smaller energy loss than does a sharp-edged entry.
3. Why does a single hydraulic grade line occur for a particular flow rate in a pipe of constant diameter and roughness?
4. Will the hydraulic grade line (in Q.3) remain the same if the flow rate is changed?
5. What type of fluid energy is used up overcoming viscous drag of the fluid on the pipe walls – i.e. in overcoming friction?
6. 'Minor losses' are usually expressed as functions of the velocity head (i.e. as $K \times V^2/2g$). Does this mean that the type of energy lost in such instances is velocity head?

7.3 The search for a frictional formula

Once it is established that frictional losses are the most important in pipeflow, it is necessary to somewhat quantify these to allow the accurate design of pipes.

This problem came to light in the nineteenth century, when the early Industrial Revolution led to a vastly increased demand for pipes to convey water to the new workshops and towns and to carry away their effluents. The requirement by clients for particular flow rates at their new factories created a serious problem for the pioneer engineers, who lacked any comprehensive understanding of energy losses in pipe flow.

Two quite distinct attacks on the problem were mounted, by the research scientists and by the practising design engineers.

The research workers had become interested in the problem around 1820, and within twenty years had produced what appeared to be a precise formula relating the head loss in pipeflow to the viscosity of the fluid, its average velocity of flow and the diameter and length of the pipe being used. However, when applied to real-life designs, much to the confusion of the scientific community, the formula (Poiseuille's equation – $H_{loss} = 32 \mu \bar{V} L/gd^2$) prove quite inaccurate. The reason for this did not become apparent for a further forty years, when Reynolds showed that fluids can adopt either the laminar or the turbulent states of flow (see Ch. 5).

The earlier researchers, using the limited laboratory equipment of the time, had restricted their experimentation to laminar flows, whereas, of course, turbulent flows dominate in most real-life applications.

As a consequence, the practising designers were forced back on their own experience and by the 1850s had produced a number of purely empirical pipeflow equations. The most important of these, the equation normally ascribed to D'Arcy, related the frictional head loss (H_{loss}) thus –

$$H_{loss} = \frac{4 \times f \times L \times \bar{V}^2}{2gd} \qquad [7.1]$$

or, substituting Q/A for $\bar{V}$,

$$H_{loss} = 64 \frac{f \times L \times Q^2}{2g \times d^5 \times \pi^2} \qquad [7.2]$$

The bulk of the terms in these equations could be measured accurately by the early design engineers and the factor f was included to make the measured head loss equate to the known length and diameter of pipe and the measured flow rate or average velocity.

Initially, the D'Arcy equation gave quite exact results. However, as engineers attempted to use it for pipe shapes, diameters and velocities, other than those from which the equation had been developed, it was soon discovered that f was far from a simple constant. Today we know that the f factor not only varies with the pipe diameter and the fluid velocity, but also with the Reynolds number of the flow (see Ch. 5).

Whilst the designers were digesting the complexity of their frictional factor and carrying out experiments on a large range of pipe materials and sizes to quantify the f term, the researchers had taken note of Reynolds' work. By the 1930s, it had been recognised that friction not only varied with Reynolds number but that its variation was quite different if the pipes were smooth or rough walled (Fig. 7.6).

Workers such as Blasius (1913) and Nikuradse (1932, 1933) produced equations for pipeflow in smooth-walled and rough-walled pipes respectively. These, whilst scientifically satisfying, proved relatively useless to practical engineers, simply because manufactured pipes are neither uniformly smooth nor rough, but a complicated mixture of the two at different points on the length of the same pipe.

Not until the late 1930s was it possible to produce mathematical expressions (the Colebrook-White transition formula, (1939)) which combined these smooth and rough pipe flow laws, and also correlated to a high accuracy with tests on commercially available pipes.

Today the work of the research scientists has been simplified to the design charts and tables published by the Hydraulics Research Station (H.R.S.) (Fig. 7.7). These allow an engineer to select the chart or table that best represents the pipe material he wishes to use and to read directly the interrelation of the four important hydraulic parameters–velocity of flow, discharge rate, hydraulic gradient and pipe diameter.

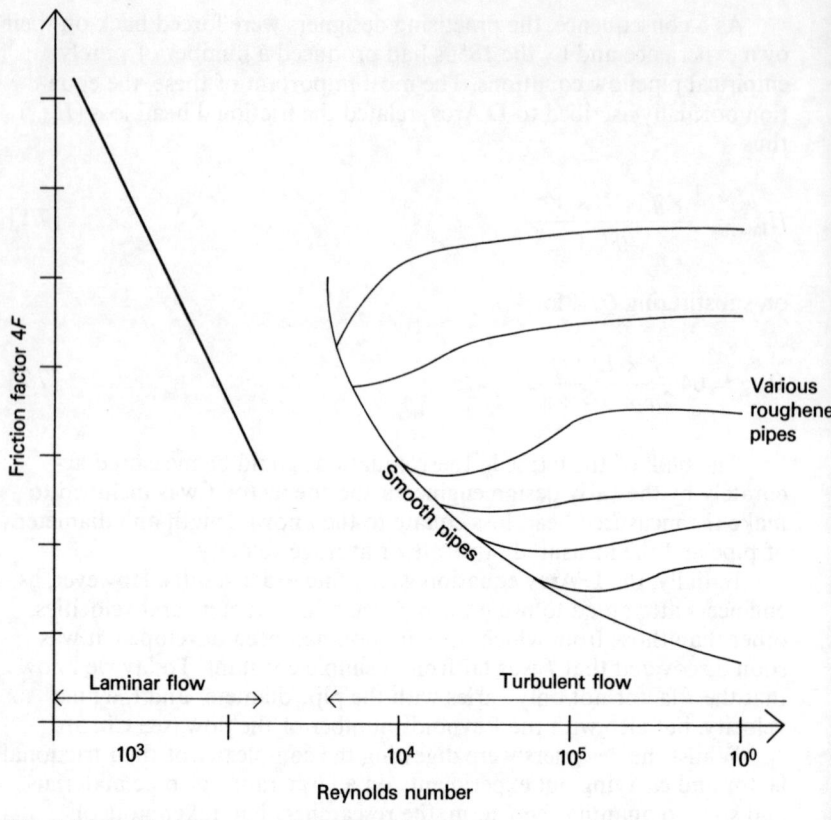

Fig. 7.6 Head losses in smooth and roughened pipes

The Colebrook-White transition formula is now recognised as the most accurate basis for hydraulic design and will give an overall accuracy of ±3 per cent. A higher level of accuracy is still beyond our knowledge and will probably remain so, until such time as the distribution of roughness in commercially produced pipes can be more precisely defined.

The older empirical formulas lost their only advantage (that of ease of application) with the publication of the Hydraulics Research Station Charts and there is no professional reason for their continued use. The D'Arcy equation however is still a useful educational tool, as it emphasises the relative effects of pipe length, flow velocity and pipe diameter on head loss. Therefore it is used in several of the following examples.

List of surfaces with roughness $k_s = 0.06$ mm

Good examples of	Galvanised iron, coated cast iron Precast concrete pipes with 'O' ring joints Spun precast concrete pipes with 'O' ring joints
Normal examples of	Wrought iron Coated steel Clayware (glazed or unglazed) with sleeve joints and 'O' ring seals Clayware (glazed or unglazed) with spigot and socket joints and 'O' ring seals – dia >150 mm Glass fibre U PVC with spigot and socket joints 'O' ring seals at 6 to 9 m intervals
Poor examples of	Uncoated steel

Discharge Q (l/s) for pipes flowing full

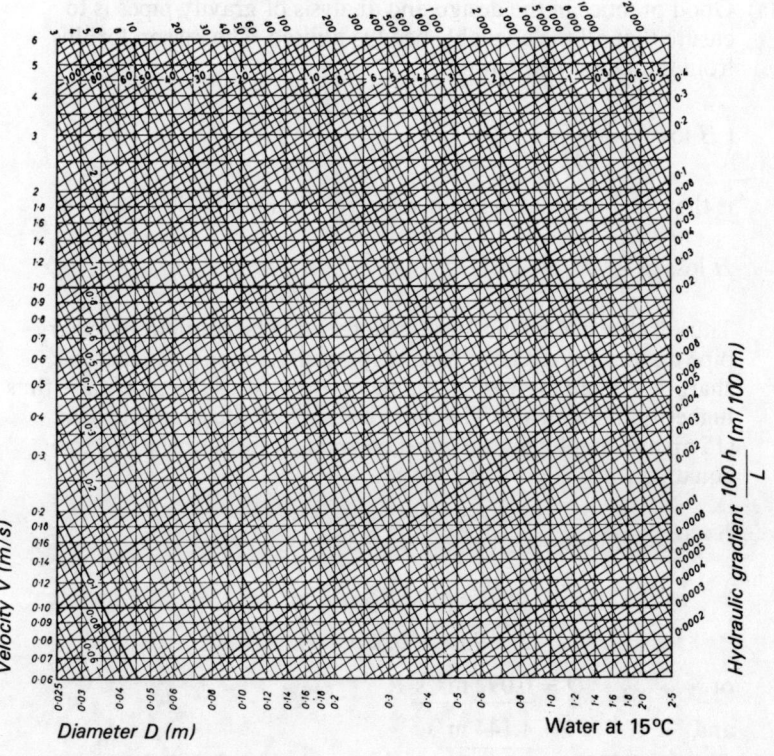

$K_s = 0.06$ mm

Fig. 7.7 Reproduced from *Charts for the Hydraulic Design of Channels and Pipes*, 4th (metric) edn, 1978, H.M.S.O., by permission of the Hydraulics Research Station

Example 7.1

A pipeline 1800 m long (pipe internal diameter 0.320 m, D'Arcy's f – 0.01) joins two reservoirs whose water surface elevations differ by 15 m.

Due to the topography, the pipe has to lie 2 m above the water level in the higher reservoir at a pipe distance of 300 m from the reservoir.

(a) Calculate the discharge through the pipe.
(b) Draw the hydraulic grade line (H.G.L.) and thus determine the pressure head at the high point of the pipe.

Solution

(a) Good practice in the design and analysis of gravity pipes is to ensure that all the available head is utilised. The reason for this, from the D'Arcy equation

$$\left(H \text{ loss} = \frac{4fLv^2}{2gd} \right)$$

is that the head loss varies inversely with the pipe diameter, i.e.

$$H \text{ loss} \propto \frac{1}{d}$$

Thus if all the available head is not used, a larger than necessary pipe diameter must have been employed. As pipe costs increase sharply with an increase in diameter, the cost of the pipeline is thus higher than necessary. Therefore utilising all the available head (15 m) and employing the most convenient form of the D'Arcy equation for this problem, we get

$$15 = H \text{ loss} = \frac{64 \cdot f \cdot L \cdot Q^2}{2gd^5 \cdot \pi^2}$$

$$= \frac{64 \times 0.01 \times 1800 \times Q^2}{2 \times 9.81 \times 0.320^5 \times \pi^2}$$

or $\qquad Q = \underline{0.092 \text{ m}^3/\text{s}}$

and $\qquad \bar{V} = \underline{1.144 \text{ m/s}}$

(b) The total energy line (T.E.L.) is, of course, the straight line joining the water surfaces of the two reservoirs.

The H.G.L. lies parallel to the T.E.L. and a distance of $V^2/2g$ metres below it, where $V^2/2g$ is a value of 0.067 m (67 mm), thus the T.E.L. and the H.G.L. are essentially coincident on a scaled drawing.

To determine the pressure head at the high spot of the pipe, which lies above the H.G.L. (and will have been identified as a syphon), it is only necessary to prepare a scale drawing of the system, and measure the vertical distance from the pipe's high point to the H.G.L. The same answer could have been obtained by applying the Bernoulli equation as in Example 6.7.

Example 7.2

The pipeline in Example 7.1 is somewhat unreal, as entry and exit losses have been ignored. More important, the pipeline lacks any control valves and its flow could neither be shut down or altered.

A more realistic example might include

an entry loss – (sharp-edged)

an exit loss   – (also sharp-edged)

on/off valves – (8 No. located at 100 m distances)

an air release valve at the syphon crest

and, a control valve to modify the flow rate, at the entry to the pipe.

The choice and available range of valves and fittings are discussed in section 7.4.

Determine:

(a) the flow rate in this more realistic pipeline;
(b) decide whether or not the inclusion of the minor losses was essential in the analysis.

Solution

(a) In this case the total available head has to be utilised overcoming the frictional losses and the turbulent eddies at the various minor loss points,

   i.e. H total $= 15$ m $= H$ loss + minor losses

   The problem is now to quantify the sum of the minor losses. Such losses are determined experimentally and appear in manufacturers' catalogues as fractions of the velocity head (Table 7.1).
   From the table, the various minor losses are

sharp entry	$0.5V^2/2g$
sharp exit	$1.0V^2/2g$
on/off valves (8 No.)	$1.6V^2/2g$
air release valve	$0.1V^2/2g$
control valve	$10.0V^2/2g$

Total minor losses $= 13.2\ V^2/2g$ metres of head

$$= \frac{13.2}{2g}\frac{(Q^2)}{A^2}\ \text{metres of head}$$

$$= 104.015Q^2\ \text{metres of head}$$

$\therefore$ as H Total $= H_L + $ minor losses

$$15 = \frac{64 \times f \times L \times Q^2}{2 \times g \times d^5 \times \pi^2} + 104.015Q^2$$

$$= 1772.98Q^2 + 104.015Q^2$$

or $Q = \underline{0.0894 \text{ m}^3/\text{s}}$

(b) As the calculated flow rate is almost identical to that in Example 7.1, it is obvious that the minor losses are indeed insignificant and need not have been included in the design.

In general this is always so unless the minor losses exceed 5 per cent of the frictional head loss.

Example 7.3

Two tanks in a water treatment plant have a 6.25 m difference in water levels and are joined by a 28 m long pipeline.

The first 20 m of the pipeline is of 100 mm internal diameter and the remaining 8 m is of 75 mm internal diameter, to reduce the dead-load on weak foundation materials.

The pipe entry, exit and change in pipe diameter are sharp.

Both pipes have frictional losses described by a D'Arcy f factor of 0.01.

(a) calculate the flow rate through the pipeline;
(b) decide whether or not the inclusion of minor losses in the analysis was necessary;
(c) draw accurate T.E.L. and H.G.L. for the pipeline flow.

Solution

(a) In this case, the total head is used in frictional losses in the two pipes and in the turbulent eddy losses,

i.e. H total $= H$ loss (pipe 1) $+ H$ loss (pipe 2) $+$ minor losses

The values of the minor losses are

pipe entry $-$ $0.5V_1^2/2g$
pipe exit $-$ $1.0V_2^2/2g$
pipe cross-section contraction $-$ $0.44V_2^2/2g$

where V_1 and V_2 are the velocities respectively in the 100 mm and 75 mm diameter pipes

as Q (pipe 1) $= Q$ (pipe 2)

$$V_1 = V_2\frac{A2}{A1} = 0.563 \ V_2$$

∴ the sum of the minor losses, in consistent terms is

entry   – $0.158\ V_2^2/2g$
exit   – $1.000\ V_2^2/2g$
contraction   – $0.440\ V_2^2/2g$

Total   – $1.598\ V_2^2/2g$

as H total $= \dfrac{4f\,Lv_1^2}{2gd\ 1} + \dfrac{4f\,Lv_2^2}{2gd\ 2} + \text{minor losses}$

$$6.25\ \text{m} = \frac{4 \times 0.01 \times 20 \times (0.563V_2)^2}{2 \times 9.81 \times 0.1} + \frac{4 \times 0.01 \times 8 \times V_2^2}{2 \times 9.81 \times 0.075}$$

$$+ \frac{1.598V_2^2}{2 \times 9.81}$$

Thus $V_2 = 3.826\ \text{m/s}$

and $V_1 = 2.154\ \text{m/s}$

Thus the discharge rate (Q) is

$$0.017\ \text{m}^3/\text{s}$$

(b) The minor losses totalled some 19 per cent of the total head available and thus are significant and must be included in the analysis. If they had been omitted, the calculated discharge rate would have been $0.019\ \text{m}^3/\text{s}$, an error of 12 per cent.

Minor losses would, of course, have been expected to be significant in a short pipeline with a relatively high velocity of flow.

(c) From the earlier calculations the following data are available.

Loss in sharp-edged entry	= 0.118 m of head
Velocity head (pipe 1)	= 0.236 m of head
Pressure head lost in 20 m length of pipe 1	= 1.90 m of head
Contraction loss	= 0.328 m of head
Velocity head (pipe 2)	= 0.75 m of head
pressure head loss in 8 m length of pipe 2	= 3.183 m of head
Exit loss of velocity head	= 0.75 m of head

To produce the required energy and grades lines, the procedure is:

1. To note that the sharp-edged pipe only reduces the total energy at the start of the 100 mm diameter pipe to 6.132 m.

2. At the end of the 100 mm diameter pipe, the total energy equals that at the entry to the pipe minus the pressure head used up in overcoming friction,

i.e. it is 6.132 m − 1.90 m = 4.232 m.

3. The total energy line for the first pipe simply joins these two levels.

4. The hydraulic grade line for the same pipe lies parallel to the T.E.L., but a distance of $V^2/2g$ (0.236 m) below it.

5. At the change in pipe diameter, an energy loss of 0.328 m occurs. Thus the total energy at the start of the 75 mm diameter pipe is that at the end of the 100 mm diameter pipe minus 0.328 m = 3.904 m.

6. This total energy line ends at the lower tank where it lies 0.75 m above the water level in the tank. Joining the start and end energy levels gives the required T.E.L.

7. As in the first pipe, the H.G.L. lies parallel to the T.E.L., but at a distance (equivalent to $V_2^2/2g$) of 0.75 m below it.

The only point to note in similar problems is that minor losses are abrupt and affect the T.E.L., whereas frictional losses are gradual and affect the H.G.L., from which the T.E.L.'s position can be found by adding the value of the velocity head.

Example 7.4

If, in example 7.3, it had proved necessary to install a 90° horizontal bend in the middle of pipe 1, what effect would this have had on the flow rate and the pipeline?

Solution

(a) *Flow-rate* – the 90° pipe bend is another minor loss, whose value is $1.15 V_2^2/2g$. This, converted in terms of V_2, is $0.365 V_2^2/2g$.

Added to the minor losses detailed in Example 7.3, this gives a total minor loss of $1.963 V_2^2/2g$ metres of head.

Thus the sums of the friction and turbulent eddy losses in the two pipes is

$$(0.129 V_2^2)_{\substack{\text{friction in} \\ \text{pipe 1}}} + (0.217 V_2^2)_{\substack{\text{friction in} \\ \text{pipe 2}}} + (0.1 V_2^2)_{\substack{\text{minor} \\ \text{losses}}}$$

$$= 6.25 \text{ m of total head}$$

$$\therefore \quad V_2 = 3.743 \text{ m/s}$$

and the discharge rate Q is 0.0165 m³/s

A slight and possibly insignificant reduction in the discharge rate of 2.9 per cent results.

(b) *Effect on pipeline*

The force on a 90° pipe bend has already been shown (Example 6.10) to be:

$$\sqrt{2}(p \cdot a \cdot + \rho Q V)$$

Using the numerical values already determined, this gives a force of

$$F = \left(p \cdot \times \pi \times \frac{0.1 \times 0.1}{4} + 1000 \times 0.0165 \times 3.743 \right)\sqrt{2}$$

The term p is the pressure intensity in pipe (1) after a flow length of 10 m. From the already calculated head loss values it can be seen that the pressure head at this point is

6.25 m − 0.118 m − 0.236   − 0.95 m

      entry   velocity   pressure
      loss   head   head loss

= 4.946 m

thus $\dfrac{p}{\rho g} = 4.946$ m

or, $p = \rho \times g \times 4.946 = \underline{48\,520 \text{ newtons m}^2}$

Thus the force on the pipe bend is 626.265 newtons acting outwards at an angle of 45° to the limbs of the pipe.

To ensure the stability of the pipe, it may be necessary to counteract this thrust by placing an anchor block on the outside of the pipe bend.

The effect of installing the 90° horizontal bend is thus to reduce the flow rate and to create an outward force on the pipe bend.

Example 7.5

A coated steel pipe, the average length of whose internal roughness elements is 0.06 mm, is to convey 0.010 m³/s of water over a distance of 1400 m. A head of 7 m is available across the ends of the pipe.

Determine a suitable pipe diameter and the average velocity of flow that will result.

Solution

In this case, the pipe material and its roughness are defined. Thus the appropriate H.R.S table (for $k = 0.06$ mm) (see Fig. 7.7) may be used.

The design flow rate of

0.010 m³/s = 10 litres/sec

and the available head (7 m) has to be expended over a pipe length of 1400 m.

Thus the required hydraulic gradient is

$$\frac{7}{1400} \text{ m} = 0.005$$

or 0.5 m for each 100 m length of pipe.

On a transparent sheet, laid over the H.R.S. chart, the calculated hydraulic gradient and discharge values can be drawn. Hydraulic gradient values are shown on the right-hand side of the chart and the lines run diagonally from top right to bottom left. Discharge values appear on the upper boundary of the chart and run diagonally, top left to bottom right.

Where the two values intersect, a line drawn perpendicular to the diameter scale gives the necessary pipe diameter of 0.130 m. A similar line drawn horizontally to the velocity scale gives the flow velocity (0.78 m/s) that would result from the use of this pipe diameter.

The H.R.S. charts are thus rapid and convenient to use, as well as providing a greater theoretical accuracy than the D'Arcy equation.

7.4 Pipe fittings

A variety of valves have been developed to ensure that the flow in a pipeline is always under control. Amongst the more important and common of these are

on/off or sluice valves,

control valves,

air release valves,

desilting valves.

7.4.1 Sluice valves

Sluice valves were invented in the early part of the Industrial Revolution (about 1850) and have proved so cheap and efficient that they have continued in use. Essentially, they are mechanisms for stopping or starting the flow in a pipeline by moving a circular plate across the pipe's cross-sectional area. Usually the plate is cranked down vertically by turning a horizontal wheel, which rotates a threaded shaft (see Fig. 7.8).

Such valves need considerable effort to close or to open, and can only operate quite slowly. This (see Ch. 9) is an advantage as it reduces the risk of the short-lived surge pressures, which might damage the pipeline if the flow rate is rapidly altered.

Because the valve depends for its watertightness on being able to slot into a close-fitting groove in the bottom of its seating, regular maintenance is required to ensure that water-borne silt does not clog up the slot. Another problem is that if such valves are left closed for long periods, it may prove impossible to open them as the circular plate may have become pressure-welded against the sides of its groove seating.

Sluice valves are quite inappropriate as methods for varying the flow rate in a pipeline, since only the last 10 per cent of valve closure meaningfully affects the flow rate. The earlier 90 per cent of valve closure only

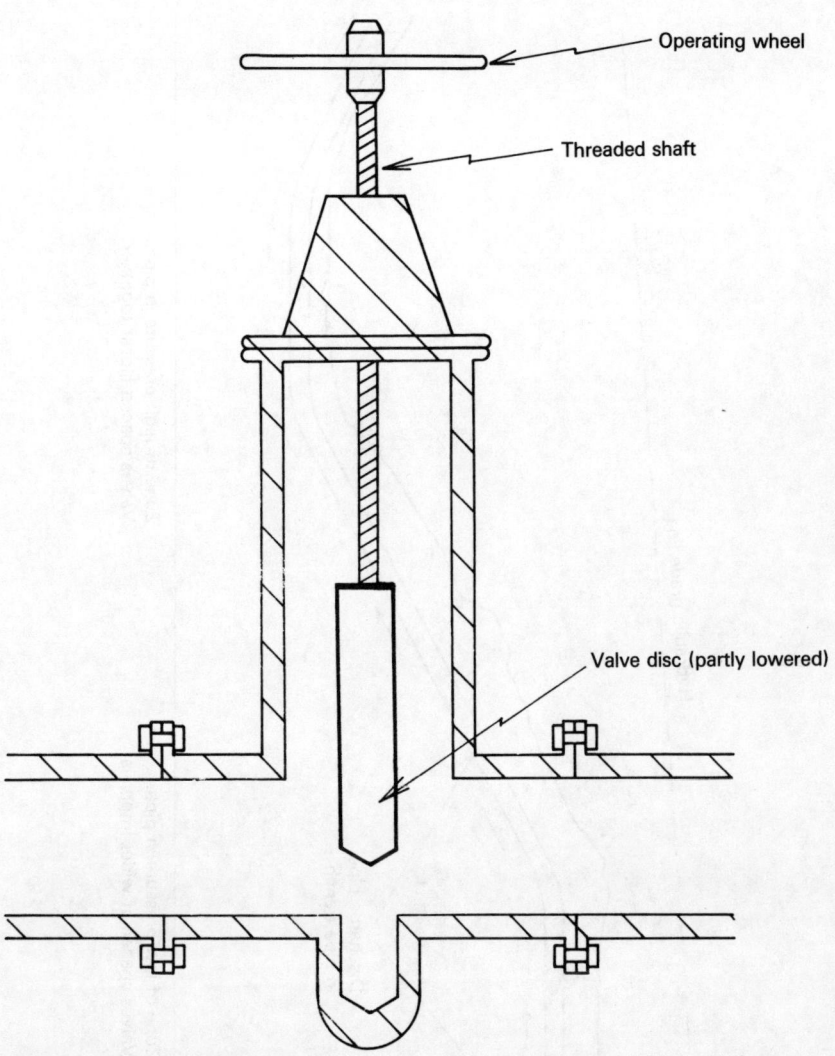

Operating wheel

Threaded shaft

Valve disc (partly lowered)

Fig. 7.8 Screw down sluice valve

forces the pipe flow through a smaller and smaller area and results in higher and higher velocities of flow. At such high velocities, the momentum forces due to the distortion of the water's streamlines can be very large and can lead to physical damage to the pipe.

Thus sluice valves are installed only to act as 'on-off' controls and give the ability to isolate sections of a pipeline for maintenance or the repair of bursts. Normal practice calls for such valves to be spaced closer together where the risk of pipe damage is greater (as for example will be the case where higher pressure heads occur) (see Fig. 7.9).

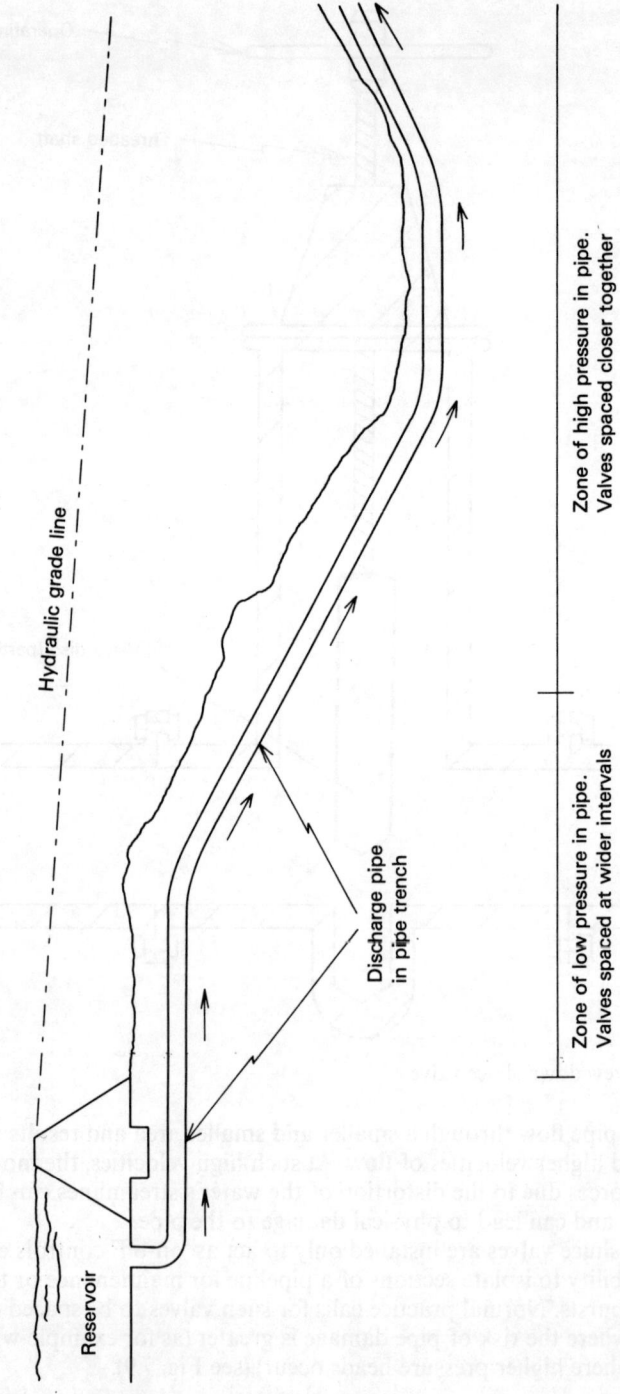

Hydraulic grade line

Discharge pipe
in pipe trench

Reservoir

Zone of low pressure in pipe.
Valves spaced at wider intervals

Zone of high pressure in pipe.
Valves spaced closer together

Fig. 7.9

7.4.2 Control valves

Control valves are essential if the flow rate in a pipeline is to be altered. Although the examples given earlier (Examples 7.1–7.4) suggests that a constant flow rate is to be expected, in fact flow rates in pipes are often altered considerably to meet variations in the daily water demand. A common example is that of the pipeline feeding domestic supplies. This will have only a relatively small discharge in the middle of the night, when demand for water will be negligible and the only likely need will be the requirements of the fire-fighting services.

To avoid the limitations of the sluice valves, control valves have to be designed to reduce the flow rate by a specified percentage for each revolution of the closing wheel. This normally necessitates variable gearing of the threaded shaft and means that control valves are more complex and much more expensive than are on-off valves. For economic reasons the installation of control valves should, thus, be the minimum necessary.

7.4.3 Air-release valves

Water tends to contain quite large volumes of dissolved air, which can escape and rise to give gas blockages at any high spots in a pipeline. These blockages – if allowed to occur – can seriously reduce the flow rate and, in extreme cases, can totally cut off the flow of water, a situation analogous to the effect of air locks in the hydraulic pipes which operate the brakes of a motor car.

To prevent the problem of gas blockages, air-release valves are installed at every high point on a pipeline.

Two quite distinct types exist, depending on whether the high point is below or above the hydraulic grade line.

Where the installation position lies below the hydraulic grade line, the pipe, of course, is full of fluid whose pressure is greater than that of the atmosphere and use can be made of the simple and cheap, ball-operated, automatic valve (Fig. 7.10).

The operation of this is quite straightforward. If air does collect in the pipe, the ball, being heavier than air, falls to the bottom of its seating, allowing the air to escape. As the air disperses and water rises into the valve seating, the ball floats up with the water and closes off the valve, to prevent loss of the water.

In an installation at the crest of a syphon, however, such a simple device would not only be ineffective but would actually increase the air blockage problem. This, of course, reflects the fact that, at a syphon crest, a pipeline contains fluid at a pressure below that of the atmosphere, and, in such circumstances, the ball would be forced to the bottom of its seating and the atmosphere would enter the pipe.

Thus a more complex, and expensive, bellows type valve is normally required for installing a syphon crest (Fig. 7.11).

These valves are normally non-automatic and have to be operated manually by maintenance personnel at fixed frequencies. To extract the air, valve B is closed, the bellows are then opened to suck out some air, valve A is then shut, valve B opened and the bellows closed to discharge

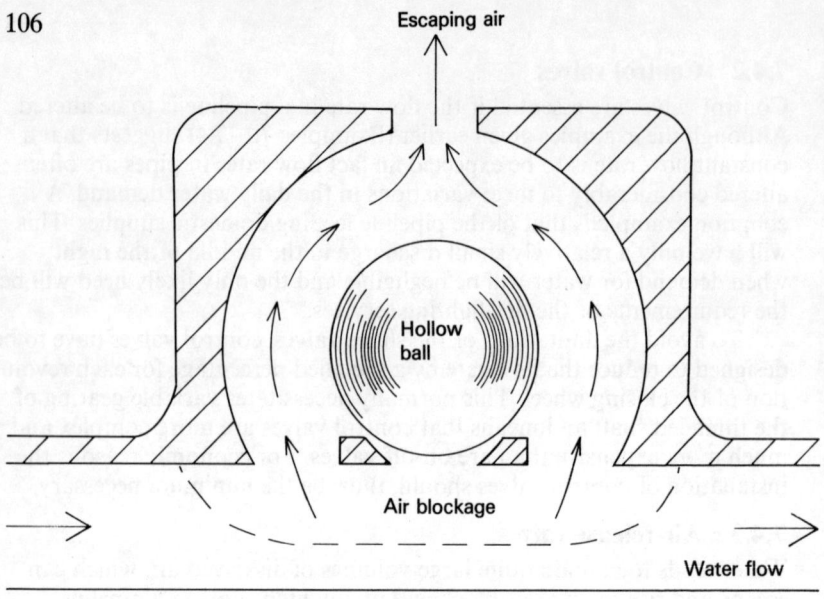

Escaping air

Hollow ball

Air blockage

Water flow

Fig. 7.10 Automatic air release valve

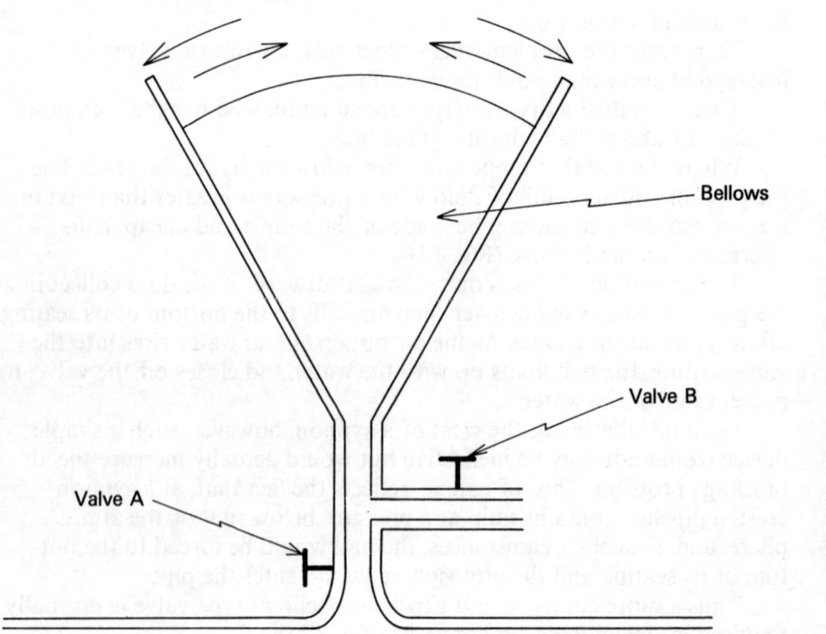

Bellows

Valve B

Valve A

Water flow

Fig. 7.11 Manual, bellows-type air release valve

the air to the atmosphere. The entire cycle is repeated until no further air exists and water is being pumped out.

7.4.4 Desilting valves

Apart from dissolved air, water invariably carries some suspended silt particles and possibly also rust flakes from the sides of the pipe. Such particles will settle wherever the flow velocity is reduced and this will happen at low spots in the pipeline. If left unchecked, the silt will gradually build up to a mass capable of blocking the pipe, and because the silt will become cemented by calcium carbonate and iron salts, from the water, it may well set into a solid mass removable only by major repair operations.

To avoid silt build-up it is usual to locate **desilting valves** (or wash-outs) at all low spots on the pipeline. These are no more than open-ended branch pipes leading to any convenient ditch or stream into which the entire flow of the pipeline can be directed for the very short period it will take to flush out any settled silt (Fig. 7.12).

Other types of valve are more appropriate for installations on pumping mains and are mentioned in a later chapter.

Valves and fittings are neither exciting nor glamorous, but without them no adequate pipeline design is possible. Thus care should be taken to consult specialised texts to understand the function and best location of the various types that are available. Probably the major reason for pipelines failing to meet client requirements is inadequate care being given to the selection and installation of valves.

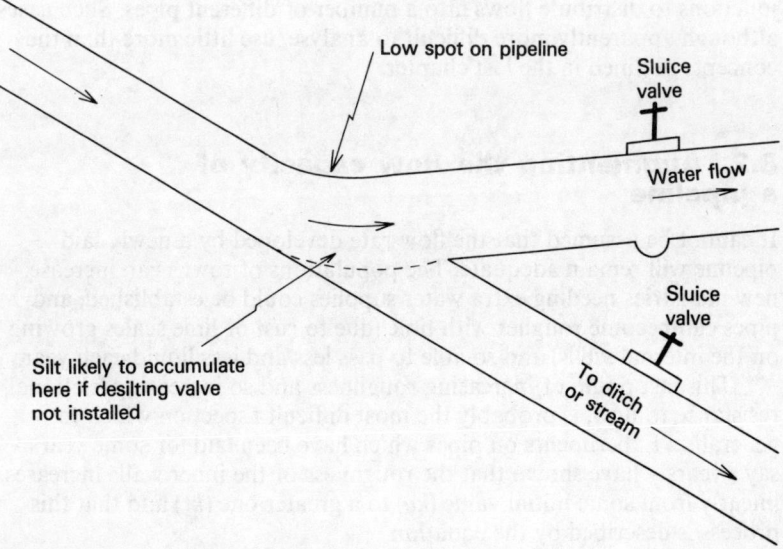

Fig. 7.12 Desilting valve

Chapter 8

More complex gravity pipes

8.1 Introduction

The pipelines described in Chapter 7 are adequate for the basic task of carrying water, or some other liquid, from a source to where it is needed.

However, more complicated pipelines are often required with junctions to distribute flows into a number of different pipes. Such cases, although apparently more difficult to analyse, use little more than the concepts outlined in the last chapter.

8.2 Augmenting the flow capacity of a pipeline

It cannot be assumed that the flow rate developed by a newly laid pipeline will remain adequate. The populations of towns can increase, new industries needing extra water supplies could be established, and pipes can become rougher with time (due to rust or lime scales growing on the internal walls) and so able to pass less and less liquid each year.

This last point, of increasing roughness and so increasing frictional resistance to flow, is probably the most difficult aspect on which to generalise. Experiments on pipes which have been laid for some years – say t years – have shown that the roughness of the inner walls increases linearly from some initial value (ko) to a greater one (kt) and that this process is described by the equation

$$k_t = k_o + Ct \qquad [8.1]$$

The factor C is a measure of the aggression of the local water in attacking, rusting or scaling the pipe walls, and only can be quantified by carrying out measurements on a pipe which has been laid for some years (see Example 8.1) in the same locality as the newly designed pipe (Fig. 8.1a, b).

If the flow rate of a gravity pipeline does become inadequate, for whatever reason, the options available for increasing the flow are limited to

(a) increasing the head across the ends of the pipe;
(b) increasing the effective diameter of the pipe

Both these solutions are obvious from a consideration of the D'Arcy equation, which can be written:

$$H_{loss} = \frac{64 \times f \times L \times Q^2}{2 \times g \times d^5 \times \pi^2}$$

$$= (\text{numerical factor}) \times Q^2$$

– to show that increasing the head gives a marked improvement in the flow rate. Similarly, it is obvious that –

$$\text{since} \quad H_{loss} = \frac{64 \times f \times L \times Q^2}{2 \times g \times \pi^2 \times d^5}$$

– for a particular flow rate, increasing the internal diameter of the pipe will reduce the head needed to establish that flow.

In many cases, increasing the head across a pipeline is either technically or economically impossible. This will be apparent if the effects of raising the water level in a reservoir are considered. The proposed higher water level could cause the flooding of adjacent roads or properties, or the increased height of dam wall could impose loads which the foundation materials cannot carry.

Thus the second option, of increasing the effective diameter of the pipeline, is often the only possible choice and the only practical method of doing this is to lay a new pipe parallel to the existing line.

On first sight, it seems both sensible and essential to parallel the entire length of the original pipe and to connect the new pipe into the source reservoir and into the lower storage tank. However, pipelines are extremely expensive, so much so that their capital cost is normally borrowed from governmental agencies or banks, and paid back, with interest, over a period of years. Thus it is important to lay only the minimum necessary length of parallel pipe to meet the water demands at the point in time.

The consequence of this is that many water supply pipelines resemble Fig. 8.2(a). Since the flow rate is steady, the flow through pipe (1) (the original inadequate capacity pipe) will equal that through pipe (2) (also original pipe) plus the flow through pipe (3) (the newly laid parallel pipe). The increased effective cross-sectional area of flow, due to pipes 2

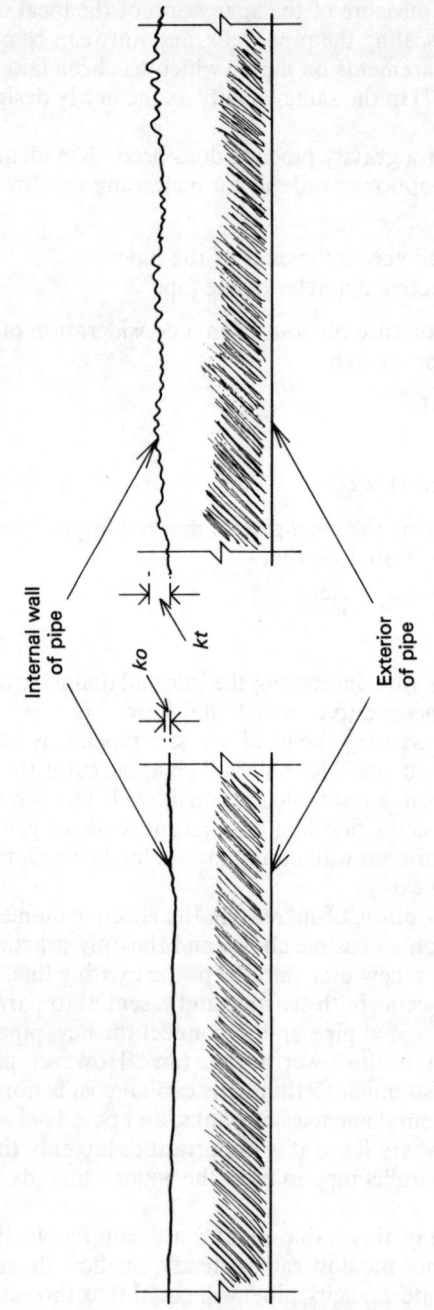

Fig. 8.1 Roughening of a pipe's internal surfaces with time. (a) Newly laid pipe (b) Pipe after *t* years of use

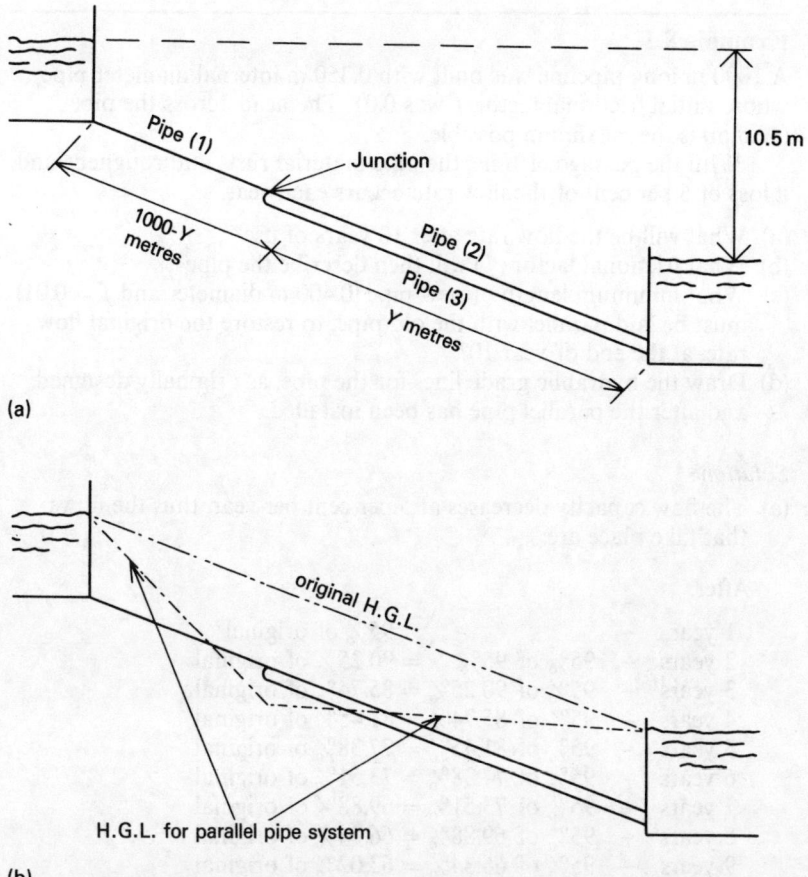

(a)

(b)

Fig. 8.2 (a) Parallel pipe system (b) Hydraulic grade lines before and after parallel pipe connected into original pipeline

and 3 sharing the discharge, will lead to a lower head loss, than previously, in that length of pipeline. Thus extra head will be available across pipe (1), than in the period before the parallel pipe was laid. This will allow the required greater discharge to be passed through pipe (1).

As the water demand increases, either because of population increase, industrial developments or further roughening of the pipe material, the length of parallel pipe can be made longer, until it totally parallels the older pipe.

In this way, only the essential capital outlay need be borne at any particular time.

The problem for a design engineer is therefore to decide exactly what length of parallel pipe need be laid to remedy a water supply shortfall.

Example 8.1

A 1000 m long pipeline was built with 0.350 m internal diameter pipe, whose initial frictional factor f was 0.01. The head across the pipe (10.5 m) is the maximum possible.

With the passage of time, the pipe material rusts and roughens and a loss of 5 per cent of the flow rate occurs each year.

(a) What will be the flow rate after 10 years of use?
(b) What frictional factor (f) will then describe the pipe?
(c) What minimum length of new pipe (0.400 m diameter and $f = 0.01$) must be laid parallel with the old pipe, to restore the original flow rate, at the end of year 10?
(d) Draw the hydraulic grade lines for the pipe, as originally designed, and after the parallel pipe has been installed.

Solution

(a) The flow capacity decreases at 5 per cent per year, thus the flows that take place are:

After:

1 year	–		95% of original
2 years	–	95% of 95%	= 90.25% of original
3 years	–	95% of 90.25%	= 85.74% of original
4 years	–	95% of 85.74%	= 81.45% of original
5 years	–	95% of 81.45%	= 77.38% of original
6 years	–	95% of 77.38%	= 73.51% of original
7 years	–	95% of 73.51%	= 69.88% of original
8 years	–	95% of 69.88%	= 66.34% of original
9 years	–	95% of 66.34%	= 63.02% of original
10 years	–	95% of 63.02%	= 59.87% of original

The original flow rate can be determined from the D'Arcy equation,

i.e. $Q_{(original)} = \underline{0.129 \text{ m}^3/\text{s}}$

and $V_{(original)} = \underline{1.343 \text{ m/s}}$

after 10 years of use, 59.87 per cent of this flow occurs

thus $Q_{(10 \text{ years})} = \underline{0.077 \text{ m}^3/\text{s}}$

and $V_{(10 \text{ years})} = \underline{0.800 \text{ m/s}}$

(b) As the only factor in the D'Arcy equation that has suffered change is the pipe roughness, the frictional factor f can be determined by inserting the ten-year flow rate in the equation,

i.e. $$H = \frac{64 \times f \times l \times Q^2}{2 \times g \times d^5 \times \pi^2}$$

or $\qquad 10.5 = \dfrac{64 \times f \times 1000 \times 0.077^2}{2 \times 9.81 \times 0.35^5 \times \pi^2}$

or $\quad f_{(10 \text{ years})} = \underline{0.028}$

(c) The layout of the pipelines is shown in Fig. 8.2(a), where the length of the parallel pipe is Y metres. The new pipe is designated 'pipe (3)' and the paralleled part of the older pipe as 'pipe (2)'.
The properties of the three pipes are thus –

pipe 1 – length – $(1000 - Y)$ metres
    – diameter – 0.350 metres
    – f – 0.028
pipe 2 – length – Y metres
    – diameter – as pipe (1)
    – f – as pipe (1)
pipe 3 – length – Y metres
    – diameter – 0.400 metres
    – f – 0.010

The head loss across pipes (2) and (3) is the same because the two pipes meet at a junction where the pressure head has the same value, and both enter the lower reservoir at the same level.

Since $\quad$ head loss$_{(\text{pipe (2))}}$ = head loss$_{(\text{pipe (3))}}$

$\therefore \qquad \dfrac{4 \times 0.028 \times (Y) \times V_2^2}{2 \times 9.81 \times 0.35} = \dfrac{4 \times 0.01 \times (Y) \times V_3^2}{2 \times 9.81 \times 0.40}$

$V_2^2 = 0.313 V_3^2$

or $\quad V_2 = 0.560 V_3$

As the flow rate is steady through the pipe network and the original flow rate of 0.129 m^3/s is to be restored,

$Q(\text{pipe 1}) = Q(\text{pipe 2}) + Q(\text{pipe 3}) = 0.129 \text{ m}^3/\text{s}$

Substituting VA for Q gives

$0.129 = \dfrac{\pi d_2^2}{4} V_2 + \dfrac{\pi d_3^2}{4} V_3$

$\qquad = \dfrac{\pi d_2^2}{4} (0.56 V_3) + \dfrac{\pi d_3^2}{4} V_3$

or $V_3 = \underline{0.721 \text{ m/s}}$

Any particular particle of water can flow either through pipe (1) and then pipe (2) or through pipe (1) and then pipe (3). In either case, the flow is powered by the total head across the pipeline. This can be expressed as

$$H_{total} = H_{loss} \text{ (pipe (1))} + H_{loss} \text{ (pipe (2))}$$

and $H_{total} = H_{loss} \text{ (pipe (1))} + H_{loss} \text{ (pipe (3))}$

substituting the calculated value of V_3 in the last equation gives

$$10.5 = \frac{4 \times 0.028 \times (1000 - Y) \times 1.343 \times 1.343}{2 \times 9.81 \times 0.35}$$

$$+ \frac{4 \times 0.01 \times (Y) \times 0.721 \times 0.721}{2 \times 9.81 \times 0.400}$$

from which

$Y = \underline{707.865 \text{ metres}}$

(d) Pipe (1) is thus 292.135 m long with a frictional factor of 0.028 and a flow velocity of 1.343 m/s. The head loss in pipe (1) is (from the D'Arcy equation) 8.594 m, as compared to 3.0 m for the same pipe with its original frictional factor of 0.01.

Thus the hydraulic grade line for the parallel pipeline is as shown on Fig. 8.2(b).

It will be noticed that one effect of installing the parallel pipe is – in this case – to expose a short length of the pipeline to pressures which are slightly below atmospheric.

8.3   Branching pipes and pipe networks

More widespread than parallel pipes are those cases where a single reservoir feeds a number of storage tanks at different levels, or where flows are distributed around a town via pipes which fork into branches of different lengths and pipe diameters. Indeed, in almost every modern city, water distribution pipes are deliberately laid around each block, as **ring mains**, to ensure that the failure of a single length of pipe does not cut off all the flow from an area.

The analysis of all such cases rests particularly in the continuity equation at each pipe junction. This can be rephrased as:

the inflow to a junction = the outflow from the junction

Two slightly different analytical techniques have been developed.

8.3.1   Head balance method

Head balance is particularly suitable for multiple reservoir cases (Fig. 8.3a), whilst quantity balance is useful for ring main problems (Fig. 8.3b).

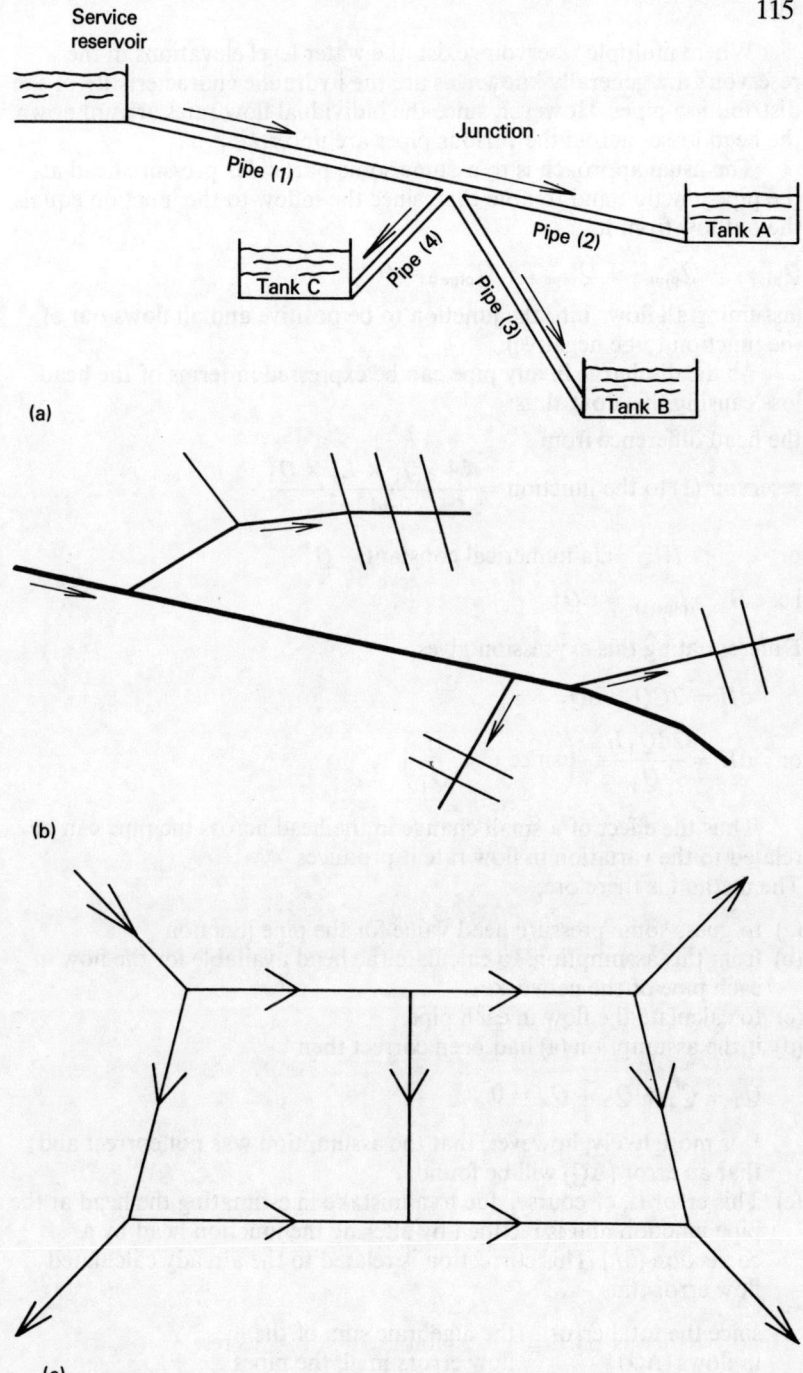

Fig. 8.3 (a) Branching pipes feeding storage tanks (b) Plan of water supply pipes in a village (c) Plan of typical city ring main

Where multiple reservoirs exist, the water level elevations in the reservoirs are generally known, as are the hydraulic characteristics of the distribution pipes. However, since the individual flow rates are unknown, the head losses across the various pipes are uncertain.

The usual approach is to assume some particular pressure head at the pipe junction and to note that, since the inflow to the junction equals the outflow from it,

$$Q_{\text{pipe 1}} + Q_{\text{pipe 2}} + Q_{\text{pipe 3}} + Q_{\text{pipe 4}} = 0$$

(assuming all flows into the junction to be positive and all flows out of the junction to be negative).

As the discharge in any pipe can be expressed in terms of the head loss causing this flow, thus:

the head difference from

$$\text{reservoir (1) to the junction} = \frac{64 \times f_1 \times L_1 \times Q_1^2}{2gd_1^5\pi^2}$$

or $\qquad H_{\text{loss}} = \text{(a numerical constant)} \quad Q_1^2$

i.e. $\quad H_{\text{loss (pipe 1)}} = CQ_1^2$

Differentiating this expression gives

$$dH = 2CQ_1 \cdot dQ_1$$

or $\quad dH = \dfrac{2dQ_1 H}{Q_1} \quad \left(\text{since } C = \dfrac{H}{Q_1^2}\right)$

Thus the effect of a small change in the head across the pipe can be related to the variation in flow rate it produces.
The method is therefore,

(a) to guess some pressure head value for the pipe junction
(b) from this assumption, to calculate the head available for the flow in each pipe of the network
(c) to calculate the flow in each pipe
(d) if the assumption (a) had been correct then

$$Q_1 + Q_2 + Q_3 + Q_4 = 0.$$

It is more likely, however, that the assumption was not correct and that an error (ΔQ) will be found.

(e) This error is, of course, due to a mistake in estimating the head at the pipe junction and is rectified by altering the junction head by a correction (dh). This correction is related to the already calculated flow error thus –

$$\text{since the total error in flows } (\Delta Q) = \text{the algebraic sum of the flow errors in all the pipes}$$

$$= dQ_1 + dQ_2 + dQ_3 + dQ_4$$

$$= \frac{\frac{1}{2}Q_1 dh}{H_1} + \frac{\frac{1}{2}Q_2 dh}{H_2}$$

$$+ \frac{\frac{1}{2}Q_3 dh}{H_3} + \frac{\frac{1}{2}Q_4 dh}{H_4}$$

$$= \frac{1}{2} dh \quad \left(\text{the sum of } \frac{Q}{H} \right)$$

or
$$dh = 2 \frac{\Delta Q}{(\text{sum of } Q/H)}$$

An illustrative example will make the procedure more apparent.

Example 8.2

A reservoir feeds two storage tanks. The source reservoir has a water level of 50 m above datum and is drained by a 300 mm internal diameter pipe 800 m long. This pipe enters a junction from which two 300 mm internal diameter pipes emerge. One is 1200 m long and leads to a reservoir whose water level is at 30 m above datum. The other is 1000 m long and drains to a reservoir whose water surface is at 25 m above datum (see Fig. 8.4).

All pipes have f factors of 0.01.

What flows enter the two storage tanks?

Solution

Any value for the head at the pipe junction can be assumed. No great care need be taken over this, since the method will easily rectify any error. An initial value of 37 m is guessed and,

the head available across: pipe 1 = 13 m   (i.e. 50 − 37 m)

the head available across: pipe 2 = 7 m   (i.e. 37 − 30 m)

the head available across: pipe 3 = 12 m   (i.e. 37 − 25 m)

Thus the flows in the three pipes are

pipe 1 = +0.109 m³/s

pipe 2 = −0.066 m³/s

pipe 3 = −0.094 m³/s

and the algebraic sum of these flows at the junction is −0.051 m³/s

This flow error (ΔQ) is obviously due to the guessed 37 m junction head being inaccurate and the head error (dh) which created it can be quantified,

from $dh = 2 \dfrac{\Delta Q}{(\text{sum of } Q/H)}$,

as   −3.984 m

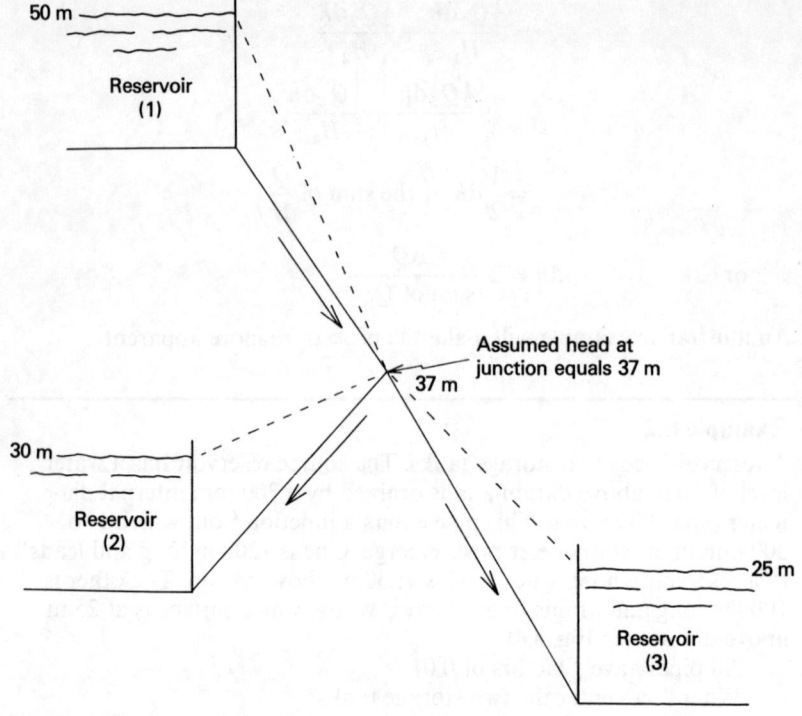

Fig. 8.4 Reservoir feeding two other reservoirs

A second attempt at the problem is then carried out using the corrected junction head of

37 m − 3.984 m = <u>33.016 m</u>

On this basis the

head available for flow in: pipe 1 = 16.984 m
head available for flow in: pipe 2 = 3.016 m
head available for flow in: pipe 3 = 8.016 m

Thus the flow in

pipe 1 = +0.125 m³/s
pipe 2 = −0.043 m³/s
pipe 3 = −0.177 m³/s

∴ ΔQ = +0.005 m³/s

and   dh = +0.320 m

In this case, the error in calculating the flow rates is obviously less significant than that of the first attempt.

A further attempt can be carried out with a junction head of 33.016 m + 0.320 m,

i.e.   33.336 m

Now, the

head available for flow in: pipe 1 = 16.664 m
head available for flow in: pipe 2 =   3.336 m
head available for flow in: pipe 3 =   8.336 m

and the flows are

in pipe 1 = +0.1238 m³/s
in pipe 2 = −0.0452 m³/s
in pipe 3 = −0.0783 m³/s

∴ ΔQ = +0.0003 m³/s

As this error is practically insignificant, the flows in the branching pipes can therefore be accepted as accurate.

___

8.3.2 Quantity balance method

The quantity balance technique is very similar. Usually the method is applied to ring main problems, where the inflows to, and the outflows from, the ring mains are known, as are the hydraulic properties of the pipes. What is normally required are the amounts and directions of flow in the individual pipes making up the ring main.

The method is:

(a) To guess the discharge values in the individual pipes, so that the flow entering a junction equals that leaving it.

(b) From these guesses, the head loss in each pipe is calculated.

(c) If the guessed flows happen to be correct, the sum of the head losses in the pipe should equal zero. The convention accepted is that all clockwise flows are positive, thus so are all head losses in such flows, and all anticlockwise flows and head losses are negative..

(d) In most cases the originally guessed flow rates will not be correct, and an error dh will be found when the head losses are summed.

(e) The flow rates are then corrected, with care being taken to ensure that the flow balance at the junctions is still maintained.

(f) The method is then applied again and a smaller error in dh should result.

(g) The procedure is halted when a practically insignificant error is found.

An illustrative example will make the procedure more apparent.

Example 8.3

The network of pipes AB, BC, CD and AD is one loop in the piping system on a steelworks site.

Inflow of 0.438 m³/s occurs at A.

Outflows of 0.130 m³/s, 0.133 m³/s and 0.175 m³/s at B, C and D respectively (see Fig. 8.5a).

The hydraulic characteristics of the four pipes are:

Pipe	Length (m)	Diameter (m)
AB	915	0.508
BC	1220	0.406
CD	915	0.406
AD	1220	0.610

All pipes have frictional factors (f) of 0.01

(a) What are the flows in the ring main?
(b) What are the pressure heads at B, C and D if that at A is 24 m of water?

Solution

(a)

Pipe	$\left(C = \dfrac{64f \cdot 1}{2 \cdot g \cdot d^5 \cdot \pi^2} \right)$	Guessed Q	$H = CQ^2$	H/Q
AB	89.32	+0.146	+1.904	13.04
BC	365.23	−0.016	−0.093	5.81
CD	273.92	−0.117	−3.750	32.05
AD	47.70	−0.292	−4.067	13.93

$dh = -6.006 \quad \Sigma H/Q = 64.83$

Note that the guessed flows are chosen to ensure that the inflow to a junction equals the outflow from it, and that the highest flows are allocated to the pipes with the lowest C values.

As the head error $\quad dh = -6.006$ m

and the sum of $\quad \dfrac{H}{Q} = 64.83,$

the initial error in guessing the flows (ΔQ) is

$$\Delta Q = \frac{dh}{2} \text{ (sum of } H/Q) = \frac{-6.006}{2 \times 64.83}$$

$$= -0.0463 \text{ m}^3/\text{s}$$

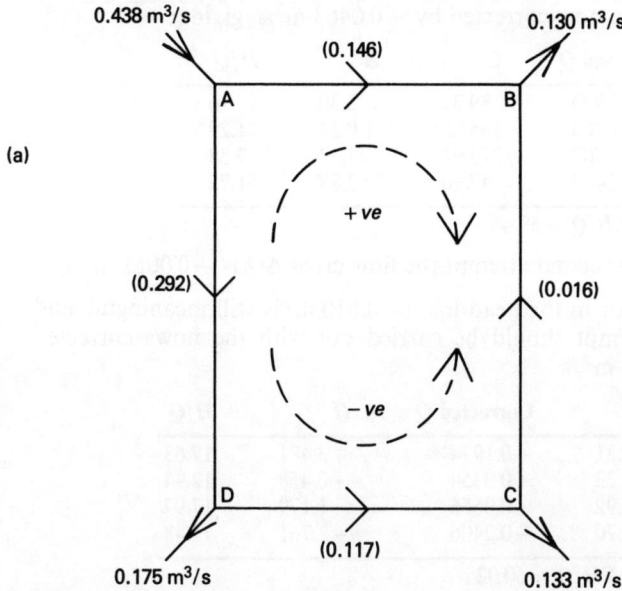

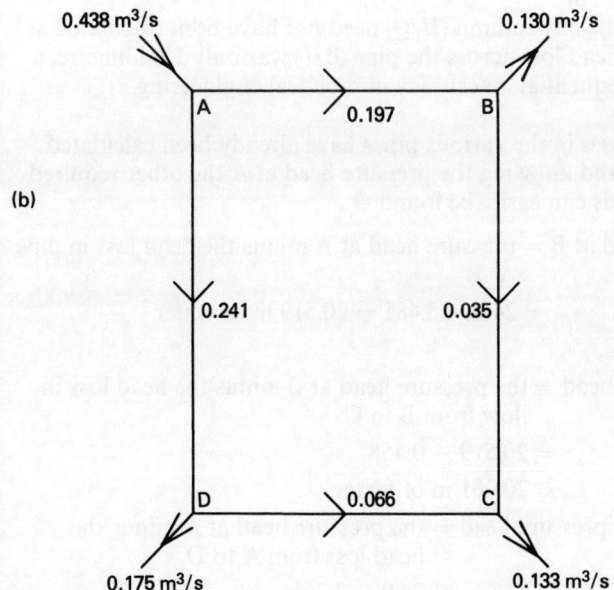

Fig. 8.5 Ring main on steelworks site. (a) Initial guessed flow directions and rates (b) Correct flows

All flows are now corrected by $+0.0463$ m^3/s, giving:

Pipe	Revised Q	C	H	H/Q
AB	+0.1923	89.32	+3.30	17.16
BC	+0.0303	365.23	+0.34	11.22
CD	−0.0707	273.92	−1.37	19.38
AD	−0.2457	47.70	−2.88	11.72

$dh = -0.61$ $\Sigma H/Q = 59.48$

Thus in this second attempt the flow error ΔQ is $\underline{-0.0051}$ m^3/s

The error in the head loss (-0.610 m) is still meaningful, and a third attempt should be carried out with the flows corrected by $+0.0051$ m^3/s

Pipe	C	Corrected Q	H	H/Q
AB	89.31	+0.1974	+3.481	17.63
BC	365.23	+0.0354	+0.458	12.94
CD	273.92	−0.0656	−1.179	17.97
AD	47.70	−0.2406	−2.761	11.48

$dh = -0.001$ $\Sigma H/Q = 60.02$

∴   the flow error $(\Delta Q) = 0.000008$ m^3/s, an obviously insignificant quantity.

The flows in the ring main can be accepted therefore as accurate (Fig. 8.5b).

In fact, the last column (H/Q) need not have been calculated as the error in head loss across the pipe (dH) was only 1 millimetre, a quite inconsequential inaccuracy in practical engineering.

(b) The head losses in the various pipes have already been calculated (H column) and knowing the pressure head at A the other required pressure heads can easily be found –

pressure head at B = pressure head at A minus the head loss in pipe A–B

$$= 24.0 - 3.481 = 20.519 \text{ m of water}$$

similarly at C,

the pressure head = the pressure head at B minus the head loss in flow from B to C

$$= 20.519 - 0.458$$

$$= 20.061 \text{ m of water}$$

and at D the pressure head = the pressure head at A minus the head loss from A to D

$$= 24.0 - 2.761$$

$$= 21.239 \text{ m of water.}$$

8.4 Computer applications

The head and quantity balance problems, given above, are relatively simple examples to illustrate these analytical methods. In practice, much more complicated cases have to be solved and very often this has to be done periodically, to evaluate the effects of changing water levels in reservoirs, or of differing inflows to ring mains.

Prior to the advent of the desk-top computer, such work occupied a large part of the time of many water supply engineers.

Now it is possible to obtain commercial computer programs to solve even the most complex cases and most engineering students, themselves, are today able to write programs to analyse quite complicated cases.

As an example, the program listed below, and written in BASIC language, will solve more difficult cases, such as Example 8.4.

Example 8.4

A centrifugal pump feeds water at a pressure head of 54.842 m into point A of the ring main shown in Fig. 8.6.

All pipes have a roughness value (k) of 0.03 mm (H.R.S tables).

Determine the flows in the various pipes and the pressure head at B by the quantity balance method.

Solution

The computer program is listed in Table 8.1.

To initiate the calculations, the following data must be input:

(a) An initial estimate of the pressure head at the pipe junction (line 00020 of program). Any value can be inserted and the errors,

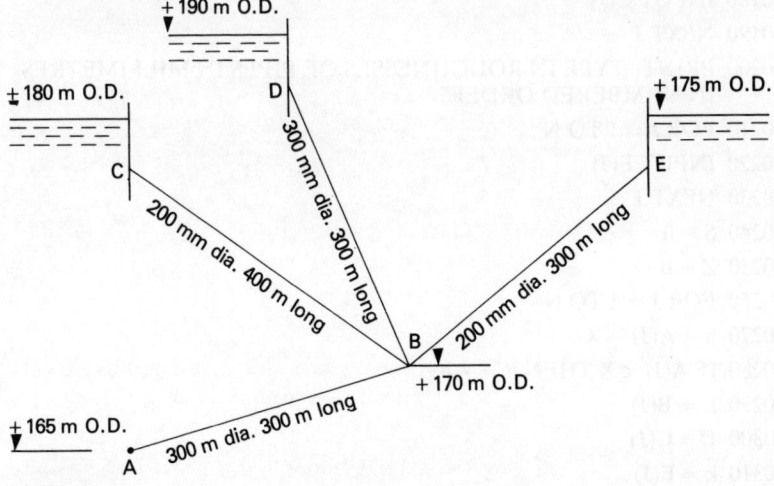

Fig. 8.6

Table 8.1 COMPUTER PROGRAM for multiple reservoir case (Example 8.4)

```
00010 PRINT "TYPE IN THE ESTIMATE OF TOTAL HEAD IN
      METRES AT JUNCTION OF PIPES"
00020 INPUT X
00030 PRINT "ESTIMATE OF HEAD AT JUNCTION ="; X, "METRES"
00040 PRINT "NUMBER THE RESERVOIRS OR PUMPED INLETS 1
      TO N"
00050 PRINT "TYPE IN THE TOTAL NUMBER OF RESERVOIRS OR
      PUMPED INLETS"
00060 INPUT N
00070 PRINT "N ="; N
00080 PRINT "TYPE IN THE WATER LEVELS IN METRES IN
      RESERVOIRS OR TOTAL HEADS"
00081 PRINT "IN METRES AT PUMPED INLETS IN NUMBERED
      ORDER"
00090 FOR J = 1 TO N
00100 INPUT A(J)
00110 NEXT J
00120 PRINT "TYPE IN THE LENGTHS OF PIPES IN METRES IN
      NUMBERED ORDER"
00130 FOR J = 1 TO N
00140 INPUT B(J)
00150 NEXT J
00160 PRINT "TYPE IN THE DIAMETERS OF PIPES IN
      MILLIMETRES IN NUMBERED ORDER"
00170 FOR J = 1 TO N
00180 INPUT C(J)
00190 NEXT J
00200 PRINT "TYPE IN ROUGHNESSES OF PIPES IN MILLIMETRES
      IN NUMBERED ORDER"
00210 FOR J = 1 TO N
00220 INPUT E(J)
00230 NEXT J
00240 S = 0
00250 Z = 0
00260 FOR J = 1 TO N
00270 h = A(J) − X
00280 IF A(J) < X THEN h = −h
00290 L = B(J)
00300 D = C(J)
00310 k = E(J)
```

```
00315  IF A(J) = X THEN 590
00320  GOSUB 540
00330  Y = Q/(2*h)
00340  IF A(J) > X OR A(J) = X THEN S = S + Q
00350  IF A(J) < X THEN S = S – Q
00360  Z = Z + Y
00370  NEXT J
00380  G = S/Z
00390  X = X + G
00400  IF G < 0 THEN G = –G
00410  IF G < 0.01 THEN 425
00420  GOTO 240
00425  PRINT "H ="; X
00430  FOR J = 1 TO N
00440  h = A(J) – X
00450  IF A(J) < X THEN h = –h
00460  L = B(J)
00470  D = C(J)
00480  k = E(J)
00490  GOSUB 540
00500  IF A(J) < X THEN Q = –Q
00510  PRINT "Q"; J; "="; Q
00520  NEXT J
00530  STOP
00540  i = h/L
00550  F = (k/(3.7 * D)) + (2.51 * 1.141 * 10 ** (–6)/
       (.001 * D * SQRT(2 * 9.80665 * .001 * D * i)))
00560  V = (–2) * SQRT(2 * 9.80665 * .001 * D * i) * LOG10(F)
00570  Q = V * PI * (.001 * D) ** 2/4
00580  RETURN
00590  Q = 0
00600  Y = 0
00610  GOTO 340
00620  END
```

(Program reproduced with the consent of the author, – Mr N. Traynor, Department of Building and Civil Engineering, Liverpool Polytechnic) resulting from the choice, will be rectified by the program. An initial choice of 170 m is probably as sensible as any other possible value.

(b) The total number of reservoirs or pumped inlets (line 00060 of program), which, in this case, is 4.

(c) The water levels or total heads of the various reservoirs in numbered order (i.e. anticlockwise) (line 00100). The pressure head at A is equivalent to 58.842 m of water, thus the total head at A (above sea-level) is 219.842 m. The total head in the reservoirs C, D and E can be read directly off Fig. 8.6 as 180 m, 190 m and 175 m respectively.

(d) The lengths of the individual pipes, in the same order as (c) (line 00140). These, of course, are 300 m, 400 m, 300 m and 300 m respectively.

(e) The pipe diameters (in millimetres) in the same order – i.e. 300, 200, 300 and 200 (line 00180).

(f) The roughnesses of the same pipes. These are derived from the Colebrook-White Formula, for the highest possible accuracy. In this case all the pipes have roughnesses of 0.03 mm (line 00220 of program).

The inputting of this data takes only a few seconds and, with a reasonably powerful computer, the required results –

Head of pipe junction   –   195.359 m

flow in pipe: (A–B)   –   0.4299 m^3/s

flow in pipe: (C–B)   –   −0.1002 m^3/s

flow in pipe: (D–B)   –   −0.1947 m^3/s

flow in pipe: (E–B)   –   −0.1350 m^3/s

are produced in a fraction of a minute.

The value of such a program is that the consequences of varying the pumped head at A can be evaluated, for as many cases as is wanted, rapidly and with a minimum demand on the engineering staff.

Thus computer programs of this type will inevitably take over the task of calculating flows in multiple reservoir and ring main cases.

Self assessment questions

1. Explain why the aging of pipes often leads to a decrease in the flow rate.

2. What do the expressions k_o and k_t mean in [8.1]? What units are they measured in? Do these expressions appear in any design chart with which you are familiar?

3. Why should laying a new pipe parallel to part of the length of an existing pipeline increase the flow rate through the pipeline?
4. Discuss the economic reasons why only part of an existing pipeline is paralleled at a time.
5. Repeat Example 8.1 with the following data:

pipeline length   –   1350 m

original diameter   –   0.250 m

original f –   0.01

head available   –   8.55 m

parallel pipe diameter   –   0.250 m

parallel pipe f –   0.01

What length of parallel pipe need be laid to restore the original flow?
6. Explain why ring mains are employed to distribute water in most urban centres.
7. List the procedures to be followed in a head balance analysis of a multiple reservoir problem.
8. What is the sign convention used in head balance analysis?
9. A reservoir A (water surface 30 m) feeds two other reservoirs (B – water surface 18 m, C – water surface 9 m) through the following pipes:

Pipe	Length (m)	Diameter (m)	f
junction – A	3000	1.000	0.014
junction – B	600	0.450	0.024
junction – C	1000	0.600	0.020

What are the flows in the three pipes?

10. List the procedures to be followed in a quantity balance analysis.
11. Why should flows in a quantity balance problem be chosen to suit the individual pipe's C values?
12. Repeat Example 8.3 with the inflow at A reduced to 0.350 m^3/s and the outflow at C reduced to 0.045 m^3/s. All other data is as shown in Example 8.3.

What are the flows in pipes AB, BC, CD and AD?

Chapter 9

Unsteady flow in pipes

9.1 The effects of varying flow rates

Until this point only steady flows, with no variation of the flow rate, have been considered.

This, of course, ignores the operational requirement to turn on, increase, reduce and halt the flow of fluid in a pipe.

Such unsteady flows persist only for fairly brief periods, but are rather more complex to analyse than are steady flows. This complexity stems from the single fact that another variable–time–has to enter the analysis.

The effect of time can be appreciated if a pipe with a fairly high velocity of flow is considered. If a valve at the end of the pipe is closed, the velocity of the fluid is reduced, and so, of course, is its velocity head. The speed with which the valve is shut governs the rate of loss of velocity head and so the rate of rise in fluid pressure (since of course fluid energy cannot be destroyed, and a loss of velocity head implies a rise in pressure head, as explained earlier). The peak value of this **surge pressure** is thus a direct function of how long the valve takes to close.

The greatly increased surge pressure at the valve can cause the pipe to burst, or more commonly, it can lead to a pressure wave that travels up the pipe, against the original flow direction, imposing greatly increased stresses on the pipe, its joints and its fittings. The usual consequence of such a wave is distorted pipe joints and leakage.

In detail, the whole process is complicated by the compressibility and inertia of most fluids. These create the situation that, at times in the development of the pressure wave, flows in two directions can exist in the same pipe.

Such unsteady flows occur even with the slow flow velocities found in domestic piping, where the noise and vibration caused by the pressure waves have led to the common name of **'water hammer'**.

9.2 Sequence of events following sudden valve closure

The sequence of events following the sudden closure of a pipeline valve is worth some detailed consideration.

If the pipeline (see Fig. 9.1), prior to any valve closure, has a steady velocity V in a pipe length L metres, immediately the valve at B is closed the fluid pressure rises. A pressure wave then moves from B to A at a velocity Vs, where Vs is the speed of sound in the fluid (1440 m/s for water). The brief interval for the wave to move from B to A is, of course,

$$\frac{L}{Vs} \text{ seconds}$$

During this first movement of the wave, water is still entering the pipe at A and the part of pipe exposed to the higher pressures of the wave has its diameter increased.

When the wave does reach A, flow can no longer enter the pipe because of the higher than normal pressures in it, and for a brief instant of time the entire water column A–B is at rest.

The wave, then, rebounds off the static fluid in the reservoir, rather like a car bouncing off a solid wall – and tracks its way back to B. In the process, this reverse wave cancels out the earlier pressure rise and when the wave finally arrives at B (at time $t = (2 \times L)/Vs$ seconds) the whole pipeline is momentarily at the pre-valve closure pressure conditions. However, the earlier high pressures (at time L/Vs seconds), together with the inertia of the fluid, has caused a delayed flow of fluid from the pipe to the reservoir and by the time $(2 \times L)/Vs$ some fluid has left the pipeline.

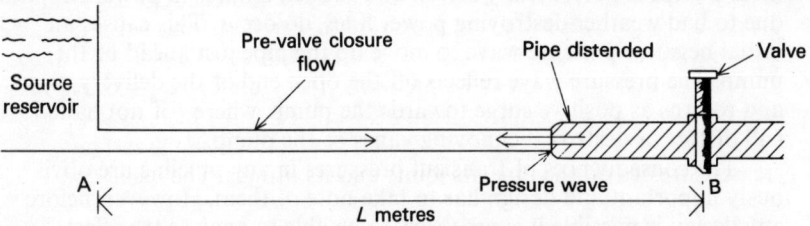

Fig. 9.1 Conditions in a pipeline immediately after a valve closure

This creates sub-normal pressure conditions and results in a second, and negative, pressure wave being generated from B back to A.

This negative wave reaches A at a time of $(3 \times L)/Vs$ seconds, when once again the fluid column is at rest. This time, however, the pipeline has sub-normal pressures throughout and, as a result, the entire pipe diameter shrinks a little.

The negative wave also bounces back from the static fluid in the reservoir and returns to B (at time $(4 \times L)/Vs$), cancelling out the sub-normal pressures in the pipe and momentarily returning the pressure and velocity conditions to those prior to valve closure.

However, the momentary sub-normal pressures in the pipe (at time $(3 \times L)/Vs$) created the conditions for extra fluid to flow from the reservoir into the pipe and this, because of the delay caused by the inertia of the fluid, only becomes significant at about time $(4 \times L)/Vs$ seconds.

Thus the sequence of positive and negative pressure waves recommences.

In theory these waves would continue indefinitely, but in reality frictional damping takes place and the waves gradually die out, and disappear completely by a time of $(20 \times L)/Vs$ seconds (Fig. 9.2).

Although the idea of positive and negative pressure waves chasing one another up and down a pipeline seems outside of everyday experience, a rather similar effect will be familiar to everyone who has travelled on a railway train.

When the train stops against the buffers in the railway station, the carriage immediately behind the engine runs forward for an extra few seconds, and compresses the coupling to the engine before coming to rest. The following carriages do likewise on one another. When the guard's van, at the end of the train, stops, the energy in the compressed couplings gives an unbalanced force, which causes each carriage in turn, to jerk backwards. This spreads as a wave down the train towards the engine. As the engine has braked, the now stretched couplings ultimately pull each carriage forwards, again as a wave moving along the train, until the entire train is again at rest. After a few cycles, friction of the wheels against the rails reduces the motion.

Almost every railway passenger has at one time or another suffered the jolting noise and discomfort of this effect.

Pumps (see Ch. 10) are also subject to a similar phenomenon. Today most pumps are electrically driven and sudden failures in power supplies, due to bad weather destroying power lines, do occur. This causes an initial negative pressure wave to move up the pipe just ahead of the pump. The pressure wave reflects off the open end of the delivery pipe and returns as positive surge towards the pump, where – if not halted – it will distort or destroy the moving vanes of the pump.

The consequences of transient pressures in any pipeline are obviously important and design has to take note of them. However, before any design is possible it is necessary to be able to analyse the effect.

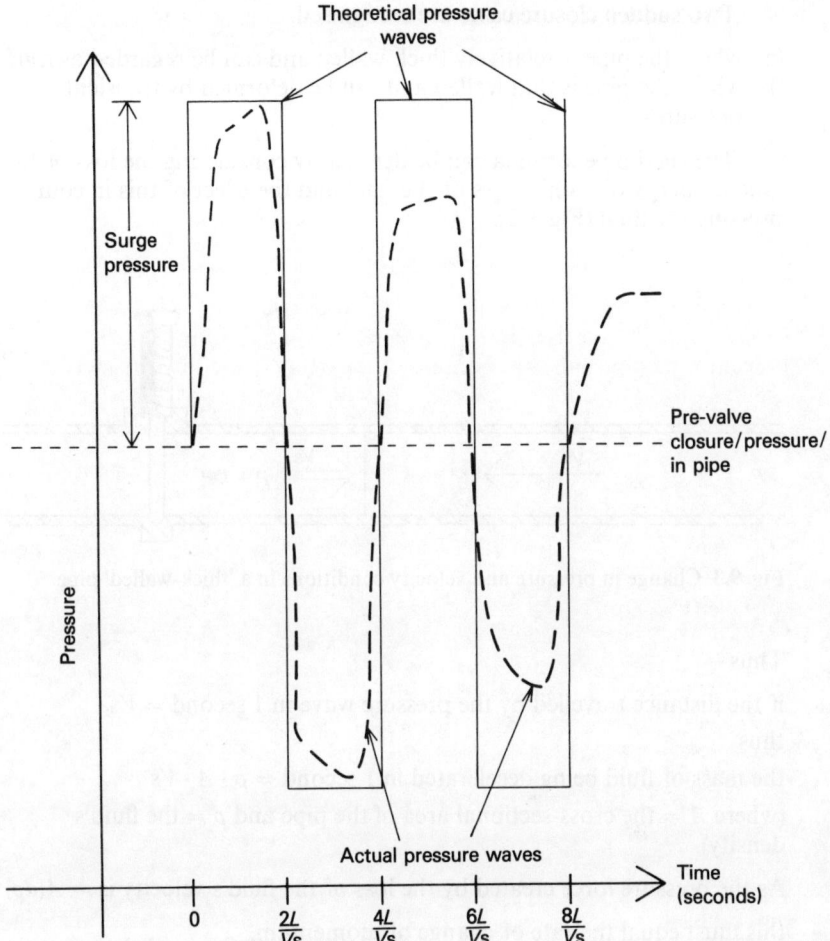

Fig. 9.2 Theoretical and actual pressure waves caused by closure of a valve

9.3 The analysis of unsteady flows

Two quite distinct methods of analysis are available, depending on whether the valve closure is 'sudden' or 'slow'.

9.3.1 Sudden valve closure

Sudden valve closure is when the valve is shut in a time of $2L/V$s seconds, or less. Obviously any analysis of this period of closure must take into account the pressure wave effects and thus the elasticity of the fluid and of the pipe material.

Two sudden closure cases are recognised:

(a) where the pipe is relatively thick walled and can be regarded as rigid;
(b) where the pipe is thin walled and can be deformed by transient pressures.

The rigid pipe formula can be derived by considering the loss of the kinetic energy of a unit mass of the fluid and the effect of this in compressing the fluid (Fig. 9.3).

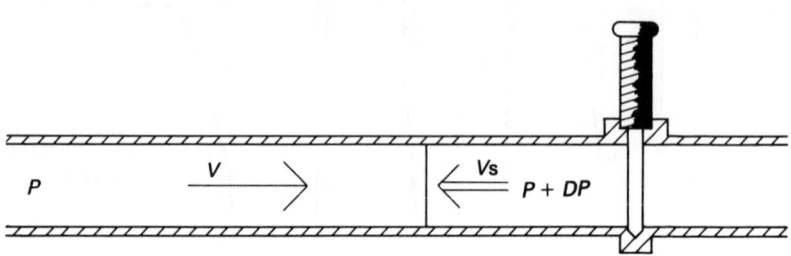

Fig. 9.3 Change in pressure and velocity conditions in a 'thick-walled' pipe

Thus –

if the distance travelled by the pressure wave in 1 second $= Vs$,

thus

the mass of fluid being decelerated in 1 second $= \rho \cdot A \cdot Vs$

(where A = the cross-sectional area of the pipe and ρ = the fluid's density).

As the pressure force created by the loss of the fluid's velocity is $= A(dp)$

this must equal the rate of change of momentum,

i.e. the increased pressure force = the loss of momentum.

$\therefore \quad A(dp) = \rho \cdot A \cdot Vs \times$ (the rate of change of the velocity)

or $\qquad dp = \rho \cdot Vs \cdot \dfrac{(dV)}{dt}$ \hfill [9.1]

and, if the original velocity is totally destroyed

$dp = \rho \cdot Vs \cdot V$ \hfill [9.2]

In the case of thin-walled pipes, the above analysis (which totally ignores the elasticity of the pipe material) is inadequate and we must argue that,

the kinetic energy lost by the fluid $=$ the strain energy gained by the fluid $+$ the strain energy gained by the pipe from which the rather more complicated formula

$$dp = \rho \cdot V \sqrt{\frac{K}{\rho}\left(\frac{1}{1 + \left(\frac{K}{E}\right)\left(\frac{D}{t}\right)}\right)} \qquad [9.3]$$

can be proved. The terms included are:

K = bulk modulus of elasticity of the fluid (i.e. $dp/d\rho/\rho$)

E = Young's modulus for the pipe material

D = the pipe's internal diameter

t = the wall thickness of the pipe

9.3.2 Slow valve closure

With the **slow closure** of a valve, the time of closure is meaningfully greater than $(2 \times L)/Vs$ seconds, and the worst of the pressure wave effects will have passed as the valve is closed. It is therefore possible – though obviously incorrect – to imagine the fluid column in the pipe as a rigid incompressible rod. Thus the effect of stopping flow at one end of the pipe can be assumed to affect the entire fluid column at the same time.

The effect of such deceleration can be analysed thus –

$$\left(\begin{array}{c}\text{the mass of} \\ \text{fluid flowing}\end{array}\right) \times (\text{the deceleration}) = \text{the increased pressure force}$$

i.e. $\left(\rho \times A \times L\right) \times \left(-\dfrac{dv}{dt}\right) = dp \times A$

or, the pressure head $(h) = \dfrac{dp}{\rho g} = \dfrac{L}{g}\left(-\dfrac{dv}{dt}\right) = -\dfrac{L}{g} \cdot \dfrac{dv}{dt}$ metres [9.4]

The above formula is based on the false assumption that the fluid column can be thought of as incompressible. As a result, if the formula is used in a case of instantaneous valve closure, the ridiculous answer of an infinitely high rise in pressure results.

In valve closures which are slower than instantaneous, but still within the period of time when pressure wave effects persist, exaggerated and inaccurate values of pressure rise result. Only when the time of valve closure equals or is greater than $20L/Vs$ seconds does the slow closure formula agree with experimental data.

Thus the slow closure formula should be applied with the care that its imprecise theoretical basis suggests.

This is a good example of the relative inaccuracy of hydrodynamic analysis, mentioned in Chapter 5. The slow closure formula as derived empirically from the results of laboratory experiments and gives excellent and accurate results for the conditions (i.e. quite slow valve closures) on which it was based. However, when applied to any more rapid closure cases an error appears and this increases as the time of valve closure becomes shorter.

Example 9.1

(a) Derive an equation relating the speed of a surge wave, caused by 'sudden' valve closure in a rigid pipe, to the bulk modulus (K) of the fluid being piped.

(b) Thus determine the speeds of surge waves for water ($\rho = 1000$ kg/m³), glycerin ($\rho = 1262$ kg/m³) and crude oil ($\rho = 800$ kg/m³).
The bulk moduli for these fluids are 216×10^9 N/m², 4.34×10^9 N/m², and 1.38×10^9 N/m² respectively.

(c) If these fluids had been flowing at 1.200 m/s in pipes of 0.300 m internal diameter, what pressure rises would have occurred?

Solution

(a) The rigid pipe formula is

$$dp = \rho \times Vs \times V \tag{9.3}$$

and was derived from the effect of compressing a particular mass of fluid.

Initially this mass was the

initial fluid density × its volume $= \rho \times (\text{volume}) = \rho \times (A \times L)$
$$= \rho \times (A \times V)$$

(where A = the cross-sectional area of flow and V is the average velocity of flow.)

As this same mass is compressed by the passage of the surge wave, it then equals

the increase in fluid density × the new volume of the fluid,

i.e. $d\rho \times A \times \text{new length} = d\rho \times A \times Vs$

$\therefore$ $\rho \times V = d\rho \times Vs$

or $$V = \left(\frac{d\rho}{\rho}\right) \cdot Vs$$

substituting this value in [9.2] gives

$$dp = \rho \times Vs \times \left(\frac{d\rho}{\rho}\right) \cdot Vs$$

or

$$Vs^2 = \frac{dp}{d\rho} = \frac{\text{the rise in fluid pressure}}{\text{the increase in fluid density}}$$

since $K = $ by definition $\dfrac{dp}{d\rho/\rho}$

$\therefore \quad Vs = \sqrt{\dfrac{K}{\rho}}$

(b) Substituting the values given for K and ρ, the required surge wave speeds are –

for water $\quad = Vs = 1470$ m/s

for glycerin $\quad = Vs = 1850$ m/s

for crude oil $= Vs = 1310$ m/s

(c) The increase in pressure due to the sudden valve closure is

$dp = \rho \times Vs \times$ velocity change,

and for water (at the flow velocity given)

$p = 1000 \times 1470 \times (1.2 - 0) = \underline{1764\,\text{kN/m}^2}$

similarly for glycerin,

$p = 1262 \times 1850 \times (1.2 - 0) = \underline{2801.64\,\text{kN/m}^2}$

and, for crude oil

$p = 800 \times 1310 \times (1.2 - 0) = \underline{1257.6\,\text{kN/m}^2}$

Example 9.2

Water in a supply pipeline has an initial average velocity of flow of 1.5 m/s. What pressure rise will result from a sudden closure of the valve at the end of the pipe? (assume the bulk modulus of water to be 2.16×10^9 N/m^2).

Solution

No mention is made of deformation of the pipe walls or of the Young's modulus of the pipe material. Thus it will be assumed that this pipe is thick walled.

Therefore the pressure rise is described by

$dp = \rho \cdot Vs \cdot V$

$\quad = \rho \times V\sqrt{K/\rho} = V\sqrt{K \cdot \rho}$

$\quad = 1.5 \times \sqrt{(2.16 \times 10^9 \times 10^3)}$

$\quad = \underline{2204.5\,\text{kN/m}^2}$

Example 9.3

A steel pipe (1.3 m internal diameter, 10 mm thick walls, and 2500 m long) carries water at a velocity of 2.0 m/s. The valve on the end of the pipe can be closed in 2.0 seconds. What pressure rise would result from such a speed of closure? (take Young's modulus (E) for steel as 207×10^9 N/m^2) and the bulk modulus of water (K) as 2.16×10^9.

Solution

First check if the valve closure is sudden (i.e. with a closure time of, at the most, $2L/Vs$ seconds).

As the speed of a surge wave in water, at normal temperatures, is 1440 m/s

$$\text{then} \quad t = \frac{2 \times 2500}{1440} \text{ secs} = 3.47 \text{ seconds}$$

The valve closure rate is faster than this, thus it must be 'sudden'.

As details of the Young's modulus and thickness of the pipe material are given, it can be assumed that this is a thin-walled pipe and that the formula

$$dp = \rho \cdot V \sqrt{\frac{K}{\rho}\left(\frac{1}{1 + \left(\dfrac{K}{E}\right)\left(\dfrac{D}{t}\right)}\right)}$$

must be used,

$$\text{i.e.} \quad p = 1000 \times 2.0 \sqrt{\frac{2.16 \times 10^9}{10^3}\left(1 + \frac{1}{\left(\dfrac{2.16}{207}\right)\left(\dfrac{1.3}{0.01}\right)}\right)}$$

$$= 1938.66 \text{ kN/m}^2$$

Example 9.4

If the valve in Example 9.3 is replaced by a slower acting valve, which closes in 35 seconds, what pressure head rise would you expect at the valve and at the mid-point of the pipe?

Solution

$$\text{As} \quad \frac{2 \times L}{Vs} = 3.47 \text{ seconds}$$

this valve closure must be 'slow'.

In fact, the time of closure is greater than $(20 \times L)/Vs$ seconds and so the incompressible theory equation [9.4] can be used with confidence, i.e.

$$\binom{\text{the extra pressure head,}}{\text{due to the valve closure}} = -\frac{L}{g} \cdot \frac{dv}{dt}$$

or $h = -\dfrac{2500}{9.81} \times \dfrac{(2.0 - 0)}{35}$

$= 14.56$ metres of water at the valve

or $= 142.83$ kN/m^2

This value is not only lower than that calculated in the preceding example, but is also affected by the length of pipe being considered. Thus the surge pressure in this particular case must reduce linearly towards the source reservoir, as shown in Fig. 9.4.

Example 9.5

A cast iron pipe has internal and external diameters of 150 mm and 180 mm. The safe pressure the pipe can take is 1.72×10^6 N/m^2.

(a) What is the maximum water flow the pipe can pass?
(b) How would this maximum value be affected, if corrosion reduces the pipe's wall thickness by 0.25 mm/year?

(assume K for water $= 2.07 \times 10^9$ N/m^2
E for cast iron $= 117 \times 10^9$ N/m^2)

Solution

(a) Enough data has been given to allow use of the thin-walled formula ([9.3])

Therefore the pressure rise is

$$dp = \rho \cdot v \sqrt{\dfrac{K}{\rho}\left(\dfrac{1}{1 + \left(\dfrac{K}{E}\right)\left(\dfrac{D}{t}\right)}\right)}$$

or $1.72 \times 10^6 = 10^3 \times V \sqrt{\dfrac{2.07 \times 10^9}{10^3} \dfrac{1}{1 + \left(\dfrac{2.07}{117}\right)\left(\dfrac{150}{15}\right)}}$

$= 10^3 \times V \sqrt{2.07 \times 10^6 \left(\dfrac{1}{1.177}\right)}$

or $1.72 \times 10^6 = 10^3 \times V \times 1326.163$

i.e. $V = 1.297$ m/s

thus the safe discharge is

$$Q = V \times A = 1.297 \times \left(\dfrac{150}{1000}\right)^2 \times \dfrac{1}{4}$$

$= \underline{0.0229 \text{ m}^3/\text{s}}$

138

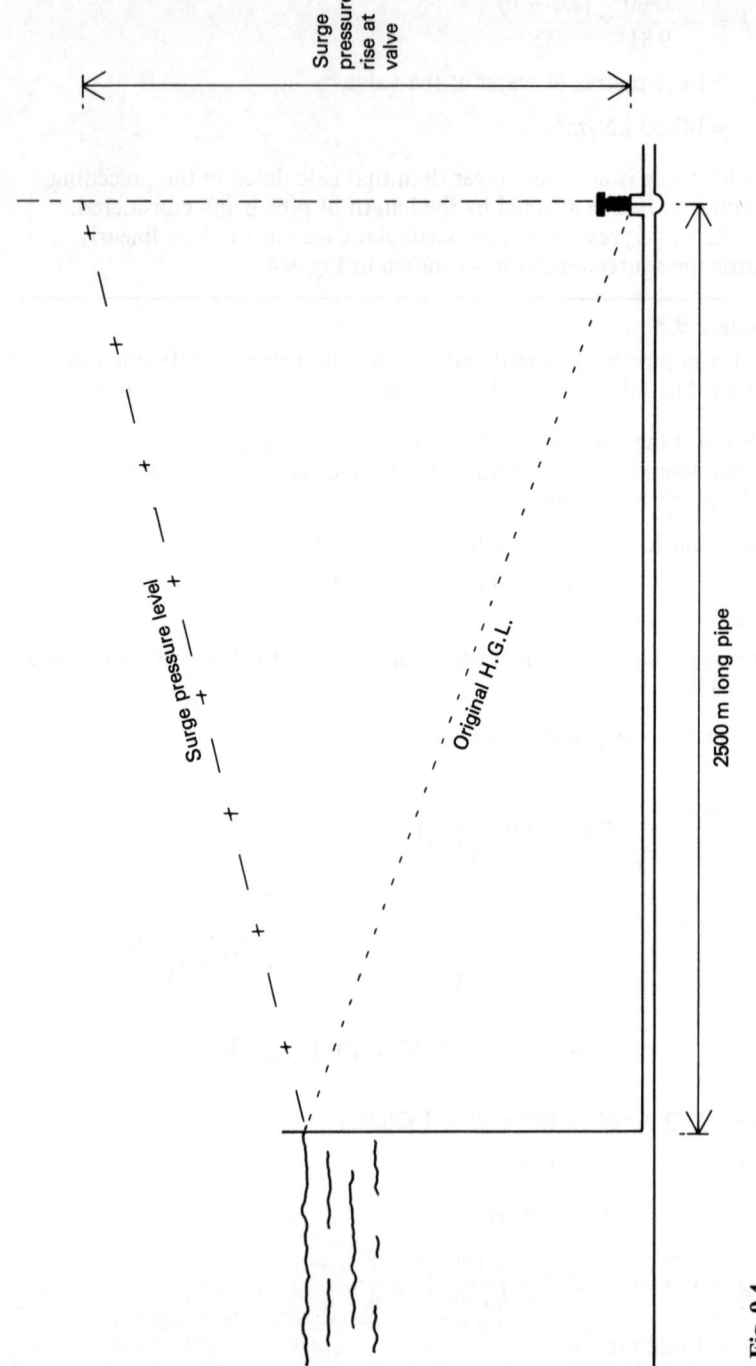

Surge pressure rise at valve

Surge pressure level

Original H.G.L.

2500 m long pipe

Fig. 9.4

(b) As the pipe wall is corroded and loses 0.25 mm of thickness each year, the safe pressure the pipe can accept will change, and so the flow rate that can safely be passed will reduce. Take any period of years that seems appropriate, determine the wall thickness of the pipe at the end of this time and repeat the above calculation.

9.4 Methods of reducing the effects of unsteady flows

The earlier section on the analysis of transient pressure variations makes it obvious that surge pressure effects are worse where pipelines are long and flow velocities are high.

Thus in domestic water supply systems the relatively short pipe lengths, together with the usually low flow velocities, allow the effects of closing taps largely to be ignored. The only precaution necessary is to ensure that no rapid closing taps are installed.

In the longer pipelines, however, design has to take note of the disruptive surge pressures. A number of alleviating methods are available.

The most obvious is to make all pipes strong enough to withstand the transient pressure fluctuations, though this is normally far too expensive, except for very short sections of pipe.

Another sensible precaution is to ensure that all valves are designed to close in as long a time as possible (see Examples 9.3 and 9.4). This, however, would not help in cases such as an electrically driven pump, whose power supply abruptly fails. In such circumstances, use can be made of automatic pressure relief valves, which open when the pressure rises to some predetermined level (Fig. 9.5a). These valves, of course, waste an appreciable portion of the flow in the pipeline and can obviously only be used where it is possible to discharge water safely from the pipe. In a built-up urban area, these valves could not normally be considered.

Flap valves – which open to permit flow in only one direction – can also be used (Fig. 9.5b), as these prevent the propagation upstream of the positive surge wave. The length of pipe between the control valve, whose closure caused the pressure surge, and the flap valve has, of course, to be armoured to withstand the surge pressure rise.

Air chambers, and their associated air compressors, are often installed immediately upstream of a pump to give protection against sudden power failure. These can be quite compact (3 m high × 2 m diameter). The air cushion at the top of the closed chamber absorbs part of the initial water hammer and the contained water is a convenient reservoir from which extra water can drain into the pipeline to dissipate the effect of the negative surge wave. Oscillation of the water

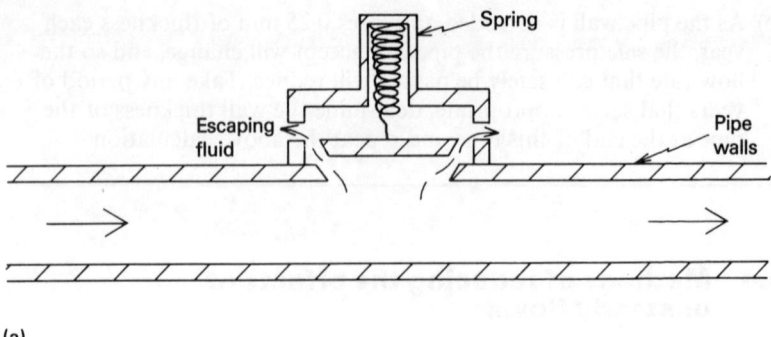

(a)

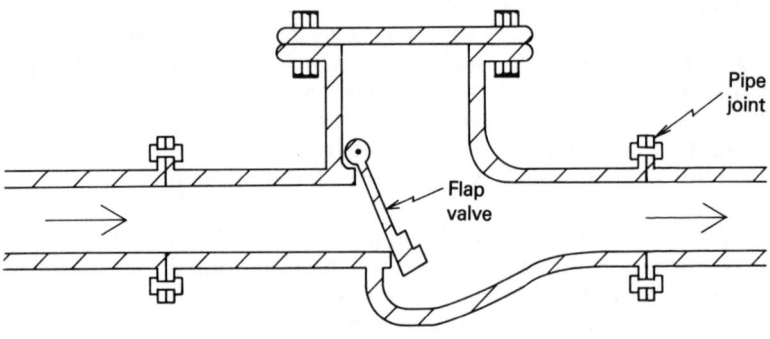

(b)

Fig. 9.5 Valves for minimising surge pressure effects. (a) Automatic pressure relief valve (b) Flap valve (can be spring loaded for faster closing)

level in the air chamber also absorbs almost all the energy of the surge wave, which is limited to travelling between the control valve and the air vessel. The air compressor may be needed to force water from the chamber to enter the pipeline faster than it would otherwise do.

The most expensive protection is that of **surge tanks**, which tend to be limited to only very large pipelines. Simply, surge tanks are open topped columns in continuity with the pipe (Fig. 9.6b).

When the flow is steady, the water level in the tank stands some distance below that of the source reservoir, due to frictional losses in the pipeline between them. When the downstream valve is closed, the resultant pressure rise causes a flow of water to enter the tank and its water level thus rises. The cycles of positive and negative pressure waves simply cause the tank's water level to fluctuate up and down until frictional losses in the pipe damp out the surge waves.

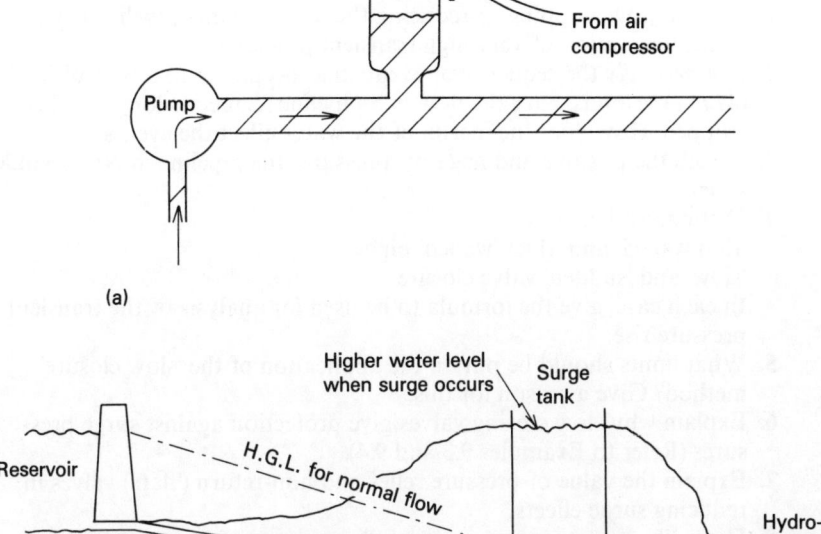

Fig. 9.6 (a) Air chamber installed on pipeline (b) Surge tank on pipeline feeding a hydroelectric generator

Normally surge tanks have to be tall enough to ensure that overflow of the water will not occur.

The pipeline between the control valve and the surge tank has, of course, to be designed to accept the full surge pressure safely.

9.5 Summary

Transient pressure effects can be much more complex than the simple cases treated here and, in most practical cases, recourse should be made to the specialised texts on water hammer analysis.

Despite the complexity of unsteady flows, the consequences of not analysing them in as great a detail as possible are too serious to ignore. An example of the possible hazards occurred at the Oigawa Power Station (Japan), where the improper closure of a valve split an 8 m length of large diameter pipe and severely buckled a further 50 m of the pipeline.

142

Self assessment questions

1. Explain how stopping or reducing the flow rate in a pipeline can lead to the generation of very high transient pressures.
2. List *precisely* the sequence of events that occurs up to a time of $(4L/Vs)$ seconds, after the flow in a pipeline (L metres long) has been stopped. How does the inertia of the water affect the events?
3. Sketch the positive and negative pressures the pipeline in No. 2 would suffer.
4. Distinguish between
   'thin walled' and 'thick walled' pipes
   'slow' and 'sudden' valve closure
   In each case, give the formula to be used for analysis of the transient pressure rise.
5. What limits should be put on the application of the 'slow closure' method? Give a reason for this.
6. Explain why slow closing valves give protection against surge pressures (Refer to Examples 9.3 and 9.4).
7. Explain the value of pressure relief and non-return ('flap') valves in reducing surge effects.
8. From library sources, prepare a fuller account of how surge tanks operate.

Chapter 10

Hydraulic machinery

10.1 Types of hydraulic machines

Amongst the more important of the machines which either use or develop hydraulic energy are –

hydraulic presses and lifts,
hydraulic ram pumps,
reciprocating pumps,
centrifugal and axial flow pumps
hydro-electric turbines of the impulse and reaction types

Although all will receive some treatment in the following sections, undoubtedly the most important hydraulic machine to a civil engineer is the centrifugal pump. Thus a deliberate bias to the selection and use of centrifugal pumping plant will be noticeable.

10.2 Hydraulic presses and lifts

These were amongst the earliest and most useful machine developments of the Industrial Revolution.

In principle they depend totally on the simplest aspect of hydrostatics, that a particular total fluid force can be given quite diverse pressure intensities by allowing it to act on surfaces of different cross-sectional area.

The most basic of these machines is the simple lift, which allows the raising of large weights by the application of much smaller forces

(Fig. 10.1). The pressure intensity caused by the force (F) on the plunger is F/a where a is the plunger's cross-sectional area and that, on the base of the ram supporting the weight, is W/A, where A is the much greater cross-sectional area of the ram.

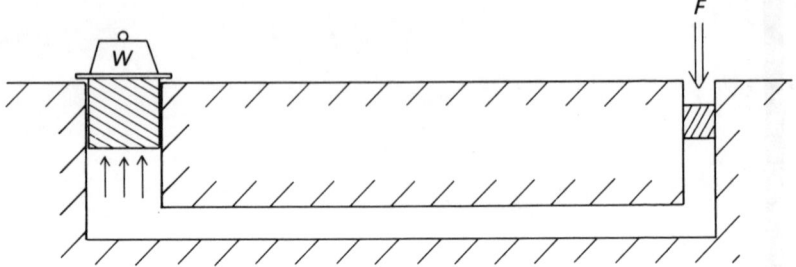

Fig. 10.1 Simple hydraulic lift

As the pressure intensities are the same everywhere throughout the fluid chamber then

$$\frac{F}{a} = \frac{W}{A}$$

or $\quad W = \dfrac{A}{a} \times F$ $\hspace{3cm}$ [10.1]

Thus the mechanical advantage of the hydraulic press is equal to the ratio of the ram and plunger areas.

For example, if the ratio of areas were 100, then a weight of one metric tonne could be raised by a force of only 1/100th of that amount.

Dockyard cranes and other fixed lifting devices are still often designed on this principle which combines the advantages of mechanical simplicity and cheapness in energy terms.

10.3 Hydraulic ram pump

The hydraulic ram pump is another product of the Industrial Revolution and the earliest type of pump which did not require an animal or human effort for its operation.

Normally ram pumps were installed to lift quite small flows to a large height above the much greater discharges in a river or stream (Fig. 10.2a). This was done by an interesting generation of unsteady flow surge pressures.

The automatic action of the ram is due to the dynamic force of water flowing into its main chamber and slamming the discharge valve A shut (Fig. 10.2b). This stops the water flow and builds up a large surge pres-

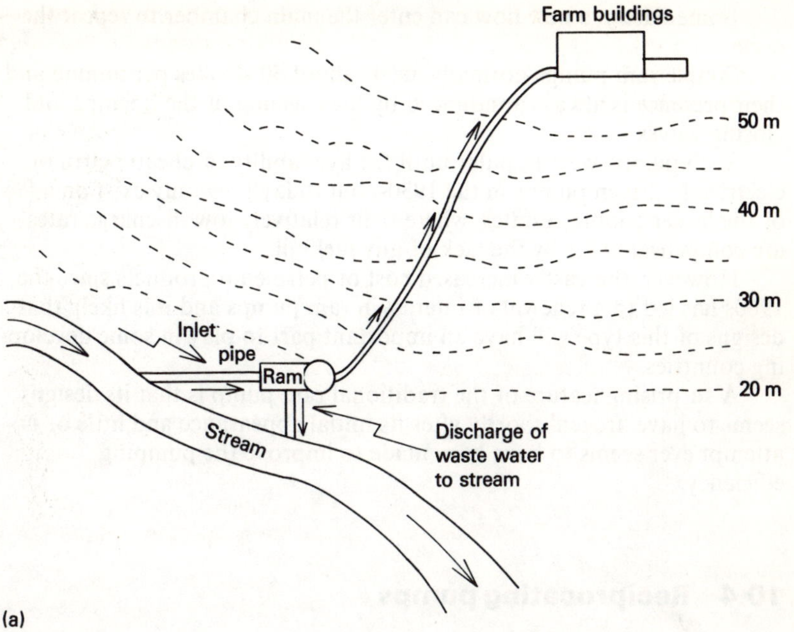

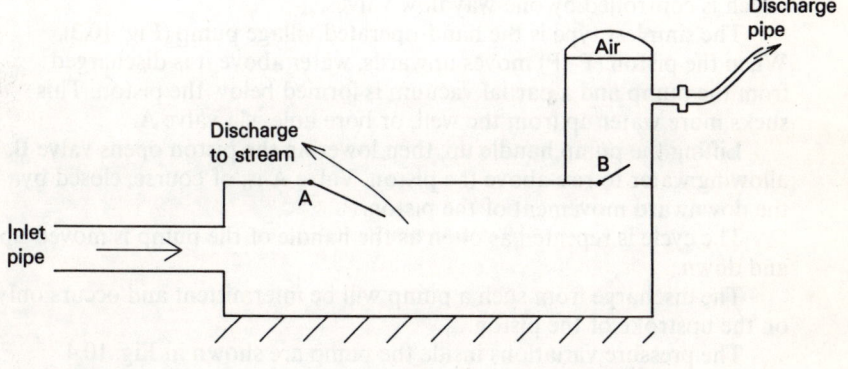

Fig. 10.2 Hydraulic ram pump. (a) Plan of ram installation (b) Section of ram pump

sure. In turn this opens valve B to the pressure vessel, where the air at the top of the vessel is compressed. This compressed air then forces some water up the discharge pipe.

When the initial pressure surge decays, valve B closes, due to its self-weight (and often because of the assistance of spring loadings), valve A opens and discharge from the main chamber back to the stream occurs.

Immediately, a new flow can enter the main chamber to repeat the cycle.

Simple ram pumps normally act at about 30 strokes per minute and their presence is always announced by the clashing of the opening and closing valves.

Such pumps were popular until the availability of cheap petrol or electrically driven pumps in the 1930s, but today they only exist on a few of the larger country estates, where their relatively low discharge rates are compensated for by the lack of any fuel bill.

However, the vastly increased cost of petroleum products since the 1960s has led to a renewal of interest in ram pumps and it is likely that designs of this type will have an important part to play in some developing countries.

A surprising feature of the traditional ram pump is that its design seems to have 'frozen' shortly after its initial appearance and little or no attempt ever seems to have been made to improve the pumping efficiency.

10.4 Reciprocating pumps

These operate by the action of a piston moving in a cylinder, access to which is controlled by one-way flow valves.

The simplest type is the hand-operated village pump (Fig. 10.3). When the piston (P–P) moves upwards, water above it is discharged from the pump and a partial vacuum is formed below the piston. This sucks more water up from the well, or bore hole, via valve A.

Lifting the pump handle up, then lowering the piston opens valve B, allowing water to rise above the piston. Valve A is, of course, closed by the downward movement of the piston.

The cycle is repeated as often as the handle of the pump is moved up and down.

The discharge from such a pump will be intermittent and occurs only on the upstroke of the piston.

The pressure variations inside the pump are shown in Fig. 10.4.

Reciprocating pumps were the earliest type widely available, and, powered by steam, allowed the construction of many of the great pioneering Victorian engineering works. To overcome pulsating flow inherent with their operation, it was common to have up to three cylinders and pistons installed in one pumping set and to operate the pistons one after another.

Amongst their advantages were reliability (due to the basic simplicity of the design), high efficiencies, and the ability to produce very high pressures. This last advantage, used to produce heads of up to 300 m of water, was gained simply by making the cylinder walls strong enough to withstand the internal pressure.

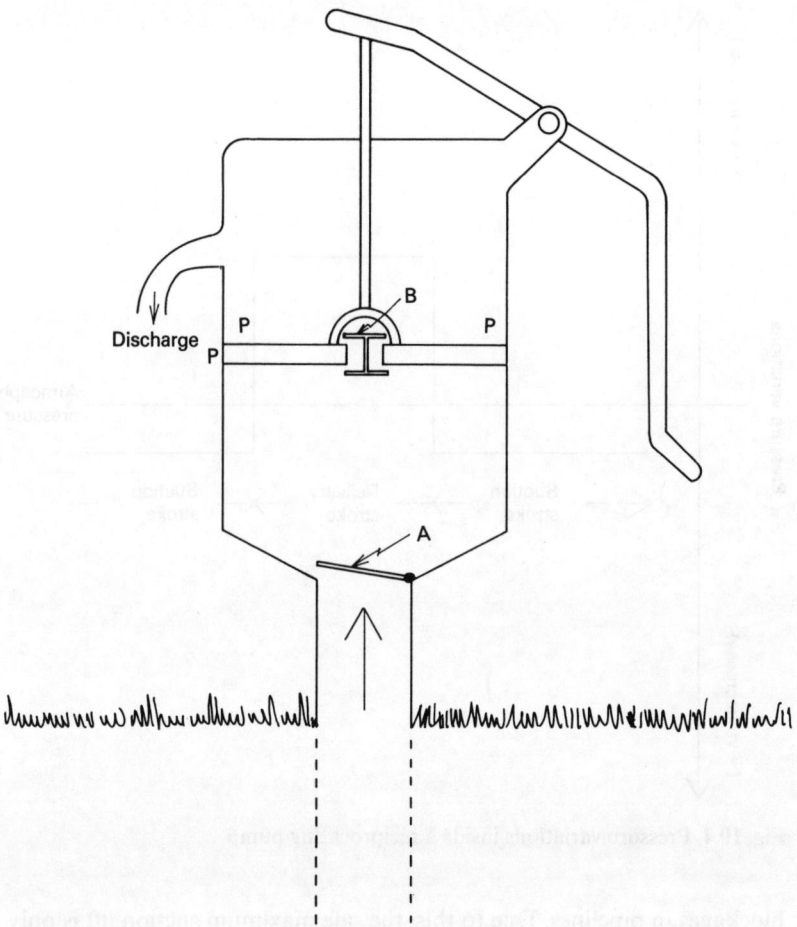

Fig. 10.3 Simple hand-operated reciprocating pump

Today, such devices have largely been replaced by more compact and cheaper centrifugal pumps. Only where precise volumes of discharge are required (as, for example, in a chemical works) or where maintenance-free running is necessary, in areas that lack specialist servicing assistance, are reciprocating pumps now common.

Apart from their necessary bulk and weight, the major factor which caused the widespread abandonment of the reciprocating pump is that a definite real-life limit exists on the suction head that can be developed. Water, like other liquids, abruptly boils to a vapour when subjected to a pressure equivalent to one negative atmosphere (i.e. equal to about 10.3 m of water head at sea level). At somewhat higher pressures, any dissolved gases in the liquid appear as bubbles in the flow, giving gas

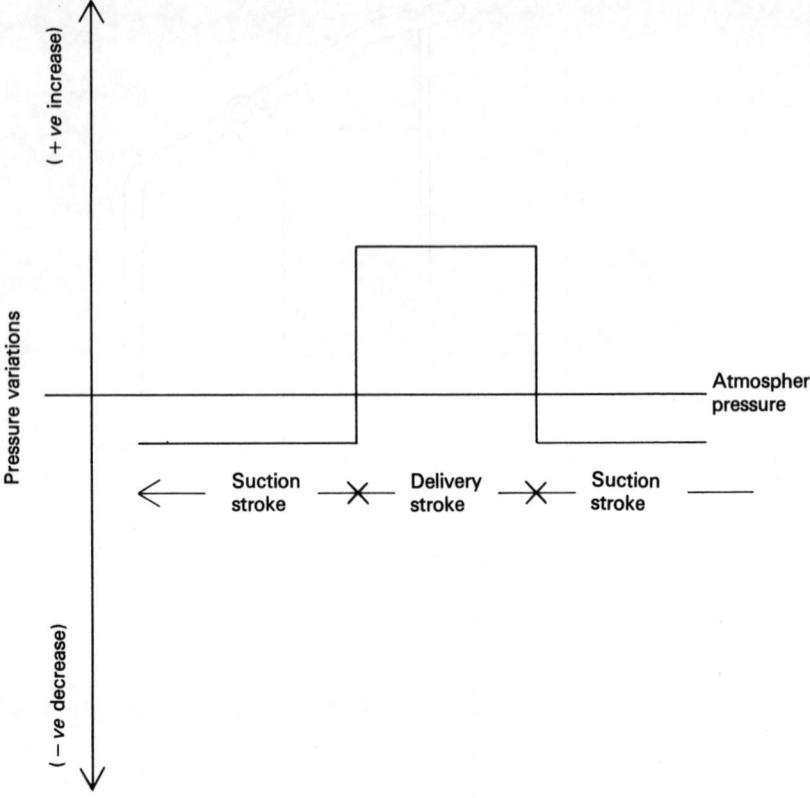

Fig. 10.4 Pressure variations inside a reciprocating pump

blockages in pipelines. Due to this, the safe maximum suction lift is only between 7 and 8 m of water. Thus reciprocating pumps cannot be installed at any greater height above the liquid they are to pump. In the dewatering of deep excavations this obviously gave rise to the practical inconvenience of moving the pumping set and steam engine down the excavation as it deepened and having to build stages on which to install it.

When centrifugal pumps appeared, their smaller mass and bulk, together with the ability to design them to operate below the liquid's surface, made the decline of the reciprocating pumps inevitable.

Sections 10.2 to 10.4 – Self assessment questions

1. Explain the operating principles of a hydraulic lift.
2. How is the mechanical advantage of a hydraulic lift obtained?
3. Does a modern hydraulic jack (for a motor car) work on the same principle?

4. How does a hydraulic ram pump operate?
5. Is it true that ram pumps tend to be low discharge and high head machines?
6. Is it possible to obtain a non-intermittent discharge from a reciprocating pump?
7. What is the pressure variation experienced in a simple reciprocating pump during one pumping cycle?
8. Why is it now possible to build hydraulic lifts, ram pumps and reciprocating pumps which require less maintenance than did the Victorian models?

10.5   Centrifugal pumps

These are the commonest of the rotodynamic pumps, so called because an internal rotating element (the 'impeller') is crucial to their operation. Over the past 50 years centrifugal pumps have been designed to cope with vastly different flow rates, a remarkable range of pumping heads, and to discharge a whole variety of fluids from air to sewage sludges. Their wide application and their ready availability from a large number of manufacturers and pump hirers have made centrifugal pumps perhaps the commonest and most useful of the civil engineer's construction tools. Thus a complete understanding of their operating principles and selection is imperative for every civil engineer.

10.5.1   Operating principles

A centrifugal pump (Fig. 10.5) consists of an inlet (or suction) pipe, a set of impeller blades, with an 'eye' connecting the inlet pipe to the impeller chamber, and a scroll case surrounding the impeller.

The impeller is a set of curved metal or hard plastic vanes, rotated by an external power source. The pump is initially 'primed,' that is, the inlet pipe up to the eye is completely filled with water or the fluid to be pumped. The motor is then switched on and the drop of the water at the pump's eye is thrown forward off the impellers. The acceleration given to this drop of water leads to a significant gain in velocity and pressure head. The removal of this initial water from the pump's eye leaves a partial vacuum behind and a further volume of water is sucked up from the inlet pipe to allow the process to continue.

As the water particles leave the impeller blades, they are collected in a scroll casing. This exists, not only to prevent the water simply spraying off like the sparks from a firework, but also to convert the now high velocity head of the water to the more useful pressure head. Pressure head is, of course, the energy form that is degraded and lost in pipe flow.

The scroll case carries out this energy conversion by a gradual increase, along the length of the chamber, in the cross-sectional area

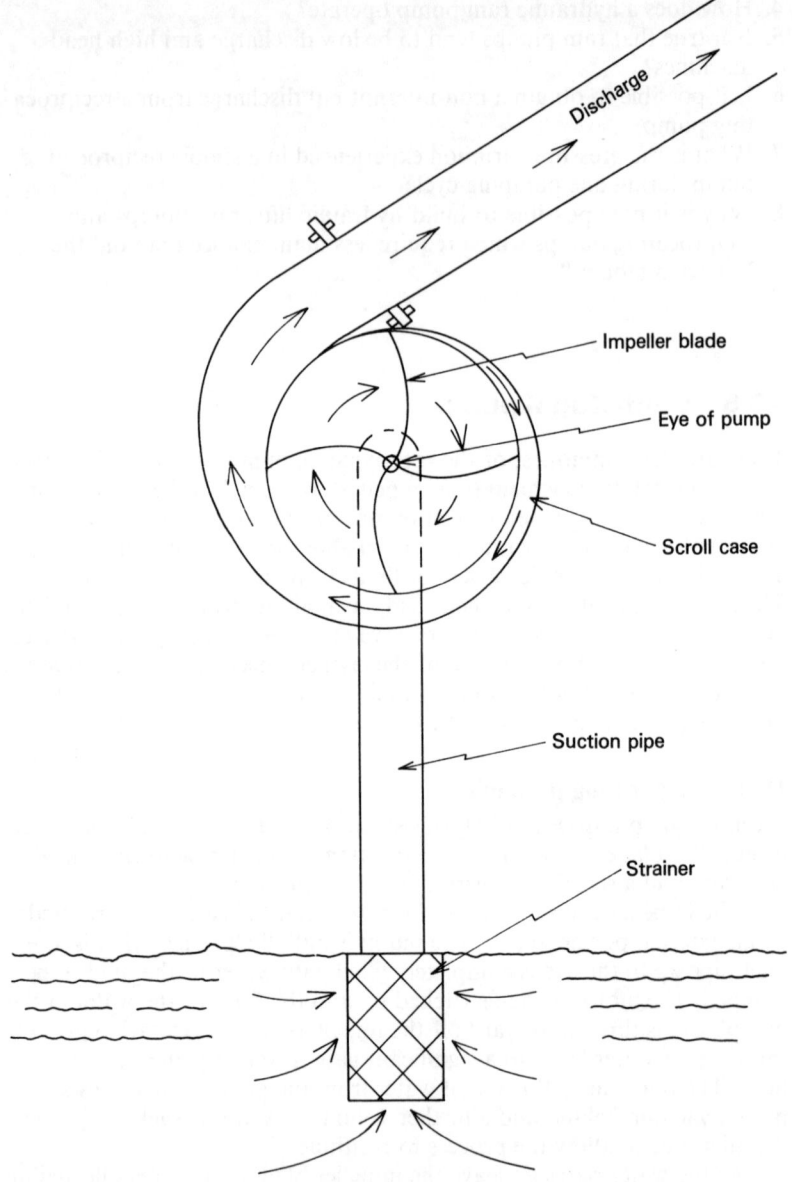

Fig. 10.5 Internal workings of a centrifugal pump

open to flow. At the scroll case's junction with the discharge pipe, all the available energy appears as pressure head and overall pumping efficiencies of 75–85 per cent are possible with good design of the expanding flow passage.

Efficiency is, of course, the ratio of

$$\frac{\text{the output pump power}}{\text{the input mechanical power}}$$

$$= \frac{\text{the weight of fluid pumped} \times \text{the head imparted to it}}{\text{the input mechanical power}}$$

$$= \frac{(\rho Q)(g)(H)}{\text{input mechanical power}}$$

$$= \frac{\rho g Q H}{\text{input power}} \qquad\qquad [10.2]$$

10.5.2 Pumping head

To make a fluid flow from one point A to another point B in a pipeline, some pressure head must be used up. In Fig. 10.6, it is therefore not

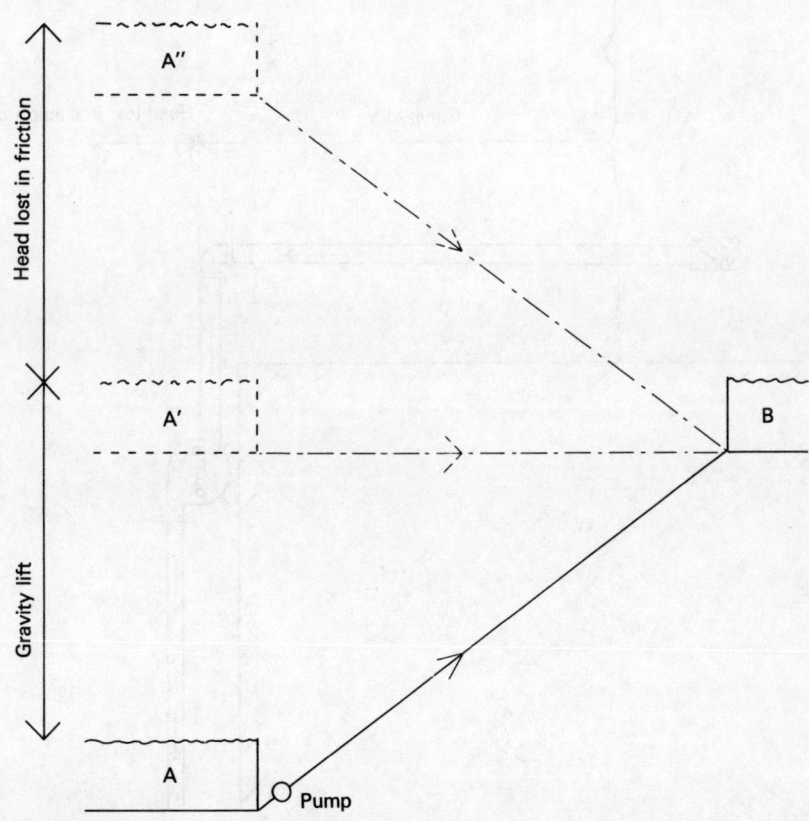

Fig. 10.6 The head delivered by a pump

152

enough to use a pump to raise the pressure head at A to that at A', since no pressure head difference would exist across the pipe A'–B and no flow could result. Instead the head at A has to be increased to that of A''.

Thus the **delivery head** of a pump equals the sum of the **gravity lift** plus the **head lost** in flow through pipe A–B, i.e.

H delivery = gravity lift + head lost in flow

Similarly, for the **suction side** of a pump (Fig. 10.7), the gravity lift + the pipe flow head losses equal the total suction head.

The total pressure head generated by the pump has to equal the sum of the suction and delivery heads, that is, the

total pumping head = suction head + delivery head

= (gravity lift + head losses) suction

+ (gravity lift + head losses) delivery   [10.3]

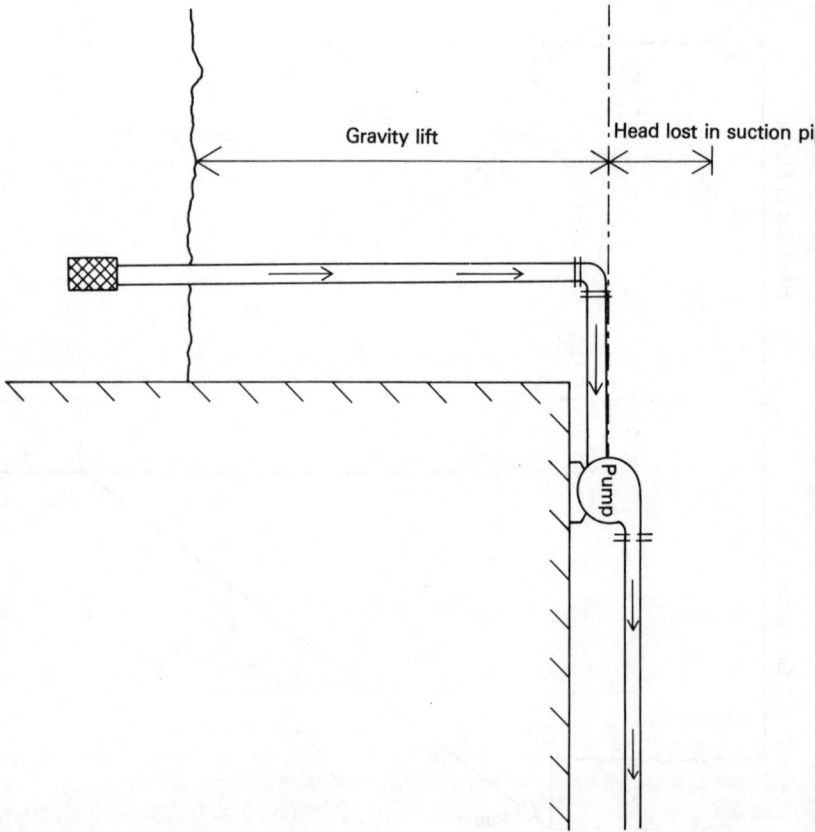

Fig. 10.7 The suction head of a pump

The head losses in both the suction and delivery pipes should (as with gravity pipes, Ch. 7) include minor head losses where these are significant. In the following examples, however, such minor losses are ignored for simplicity.

Example 10.1

A pump is connected to a 47 m length of 0.300 m internal diameter suction pipe and discharges into a 350 m long pipe (internal diameter 0.150 m). The water source is a river, whose water level is 2 m above datum, the centre line of the pump is 3.75 m higher, and the discharge pipe enters a tank whose water level is at 22.5 above datum.

What head must the pump produce if $0.100 \, \text{m}^3/\text{s}$ is to be pumped? (Assume all pipes are described by a D'Arcy's f factor of 0.01, and ignore any minor losses in the pipelines).

Solution

The pumping head = the delivery head + the suction head

The delivery head = the gravity lift + the pipe flow losses

$$= (22.5 - 5.75) \, \text{m} + \frac{64 \times f \times L \times Q^2}{2 \times g \times d^5 \times \pi^2}$$

$$= 16.75 + 152.333 \, \text{m}$$

$$= \underline{169.083 \, \text{m}}$$

Similarly, for the suction side,

the head = gravity lift + pipe flow losses

$$= 3.75 \, \text{m} + \frac{64 \times 0.01 \times 47 \times 0.1^2}{2 \times 9.81 \times 0.3^5 \times \pi^2} \, \text{m}$$

$$= 3.75 + 0.639 \, \text{m}$$

$$= 4.389 \, \text{m}$$

Thus the total head required from the pump is

$$= 169.083 \, \text{m} + 4.389 \, \text{m}$$

$$= \underline{173.472 \, \text{m}}$$

Example 10.2

The centrifugal pump in Example 10.1 is replaced by a submersible version placed 1 m below the source water's level. The delivery pipe length is necessarily increased by 47 m. All other factors remain the same.

What head must the submersible pump produce?

Solution
As before,

the total pump head = the delivery head + the suction head.

In this case, the suction head is negative, since the pump lies below the water level by 1 m. No pipe losses on the suction side occur.

Thus the suction head = -1 m

The delivery head = the gravity lift + the pipeflow losses

$$= (22.5 - 1)\,\text{m} + \frac{64 \times f \times L \times Q^2}{2 \times g \times d^5 \times \pi^2}$$

$$= 21.5 + 172.789\,\text{m}$$

$$= \underline{194.289\,\text{m}}$$

and the total head on the pump = $\underline{193.289\,\text{m}}$

Naturally the effect of increasing the length of the smaller diameter delivery pipe has led to an increase in the head the pump has to deliver.

Submersible versions are often employed in preference to normal centrifugal pumps where access to the source water is by a narrow diameter borehole or where the only dry site for a pump is so high above the source water level that an unacceptably high suction head will result.

However, in this particular case, the installation of a submersible pump seems unwarranted.

Example 10.3

To prevent the loss of drinkable water to the sea, a major water transfer scheme consists of 2 km of 0.500 m internal diameter steel pipe laid over the flood plain of the river estuary (level 2.5 m above datum); a pumphouse on the edge of the flood plain, level 5 m above datum; 3 km of 1.000 m internal diameter steel pipe laid up a steep hill (level of hill top 42 m above datum).

From the hill top the pumped water (0.350 m^3/s) drains by an open channel to another river.

(a) Determine the delivery and suction heads required of the pump.
(b) Draw accurate H.G.L.s for the operation of the pump.
(c) Comment on the use of the pipe diameters chosen (take all pipes as described by the H.R.S. tables, where $k = 0.06$ mm).

Solution
(a) For the delivery side of the pump,

the necessary head = gravity lift + pipe flow losses

$$= (42 - 5)\,\text{m} + \text{pipe flow losses}$$

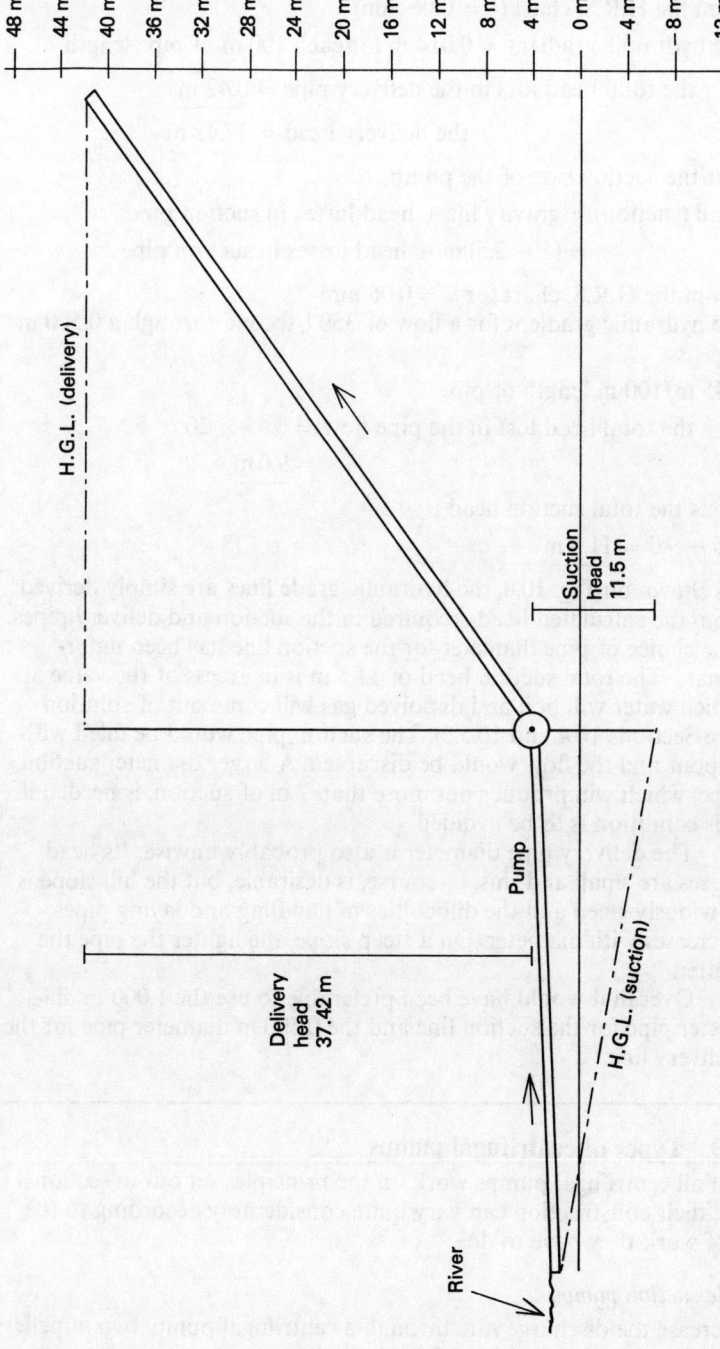

Fig. 10.8 Example 10.3 – hydraulic grade lines

48 m
44 m
40 m
36 m
32 m
28 m
24 m
20 m
16 m
12 m
8 m
4 m
0 m
-4 m
-8 m
-12 m

155

H.G.L. (delivery)

Suction head 11.5 m

Pump

H.G.L. (suction)

Delivery head 37.42 m

River

from the H.R.S. chart ($k = 0.06$ mm)

the hydraulic gradient = 0.014 m for each 100 m of pipe length

∴   the total head loss in the delivery pipe = 0.42 m

∴   the delivery head = 37.42 m

For the suction side of the pump,

head (suction) = gravity lift + head losses in suction pipe

$$= (5 - 2.5) \text{ m} + \text{head losses in suction pipe}$$

From the H.R.S. chart for $k = 0.06$ mm
the hydraulic gradient for a flow of 350 l/second through a 0.500 m pipe is

0.45 m/100 m length of pipe

∴   the total head lost in the pipe flow = 0.45×20

$$= 9.0 \text{ m}$$

Thus the total suction head is

$2.5 + 9.0 = 11.5$ m

(b) As shown on Fig. 10.8, the hydraulic grade lines are simply derived from the calculated heads required in the suction and delivery pipes.

(c) The choice of pipe diameter for the suction line has been unfortunate. The total suction head of 11.5 m is in excess of the value at which water will boil and dissolved gas will come out of solution (see Sections 10.4 and 10.5.5). The suction pipe would be filled with vapour and the flow would be disrupted. A larger diameter suction pipe, which will produce not more than 7 m of suction, is needed if this condition is to be avoided.

The delivery pipe diameter is also probably unwise. Its head losses are small and this, of course, is desirable, but the hill slope is obviously steep and the difficulties of handling and laying pipes increases with diameter. On a steep slope, the lighter the pipe the better.

Overall it would have been preferable to use the 1.000 m diameter pipe for the suction line and the 0.500 m diameter pipe for the delivery line.

10.5.3   Types of centrifugal pumps

Whilst all centrifugal pumps work on the principles set out in section 10.5.1, their construction can vary quite considerably according to the type of work they have to do.

Double suction pumps

To increase the discharge rate through a centrifugal pump, two impellers can be placed back to back, with each discharging into its own scroll case. A similar result can be obtained by allowing water to enter the impeller from both sides. Such double suction pumps are useful as they are more compact than the other alternatives.

Multistage pumps

When very high pumping heads are needed (i.e. heads in excess of 100 m), a simple centrifugal pump, of the type shown in Fig. 10.5, would have to generate a very high velocity head, which would then be converted in the scroll case to the required pressure head. To do this, the pump would either have to have an extremely fast impeller rotation or impellers of a very large diameter. As either solution would be inconvenient, multistage pumps have been developed, with a number of identical impellers all driven by the same motor shaft. The discharge from the first impeller is arranged so that this enters the eye of the second impeller and similarly with all other impellers. The consequence of this is that each impeller increases the water's pressure head by the same amount and as great a head as required can be obtained simply by adding impeller stages.

Borehole pumps

When a pump has to work in the very limited space inside a borehole, a number of impeller stages invariably have to be used to make up for the necessarily reduced impeller diameter. Often the impeller stages in such pumps are simply bolted together, allowing their addition or removal (Fig. 10.9).

If the depth to water in the borehole decreases or increases significantly, stages can thus be subtracted or added to produce the required pumping head.

Such pumps can be driven by motors at ground surface, via a shaft of suitable length, or a waterproofed motor can be mounted with the impeller stages below the water surface. This later design is termed a submersible pump and has generally superseded the shaft-driven models.

Sludge pumps

When a fluid contains a high proportion of suspended solid particles, the pump necessarily has to have impellers of a material tough enough to withstand abrasion and flow passages wide enough to avoid clogging up problems. A whole range of such pumps is available and from an external view resemble normal centrifugal pumps.

The choice of a particular centrifugal pump is fairly straightforward. The usual horizontal axis centrifugal pump is cheap and usually ideal for most civil engineering applications. It can be wheel mounted for mobility on construction sites and is available in a remarkably wide range of heads and discharges.

For wells and boreholes, the choice lies between the shaft driven and the submersible multi-stage units and the submersible design is often preferred.

158

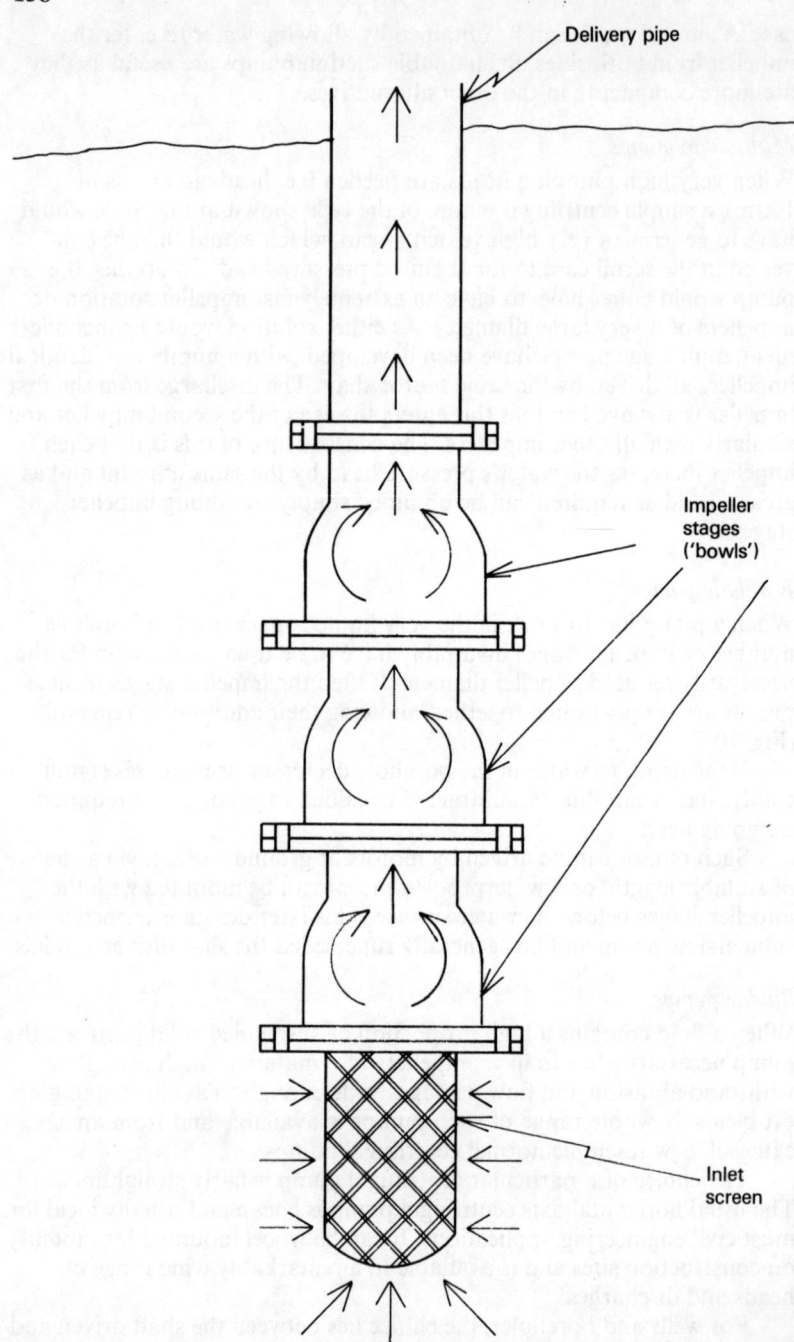

Fig. 10.9 Multi-stage borehole pump

For high-lift pumping, the multi-stage horizontal axis centrifugal pump is normally the most suitable, though submersible pumps, installed directly in pipelines, can be used and save the cost of constructing a pumping house.

Whatever pump is chosen, it is good practice either to have a stand-by unit available, except where a pump hirer is conveniently close by. This avoids lengthy and costly delays due to a mechanical failure.

10.5.4 Characteristic curves for a centrifugal pump

The properties which interest an engineer buying or hiring a pump are:

What head range does the pump have?

What discharges can it pass?

How efficient is the pump?

What power input will it need?

Obviously the first two properties are inter-related, since the output power of a pump (from [10.2] is $\rho g Q H$. Thus if the head is varied, so the discharge rate changes.

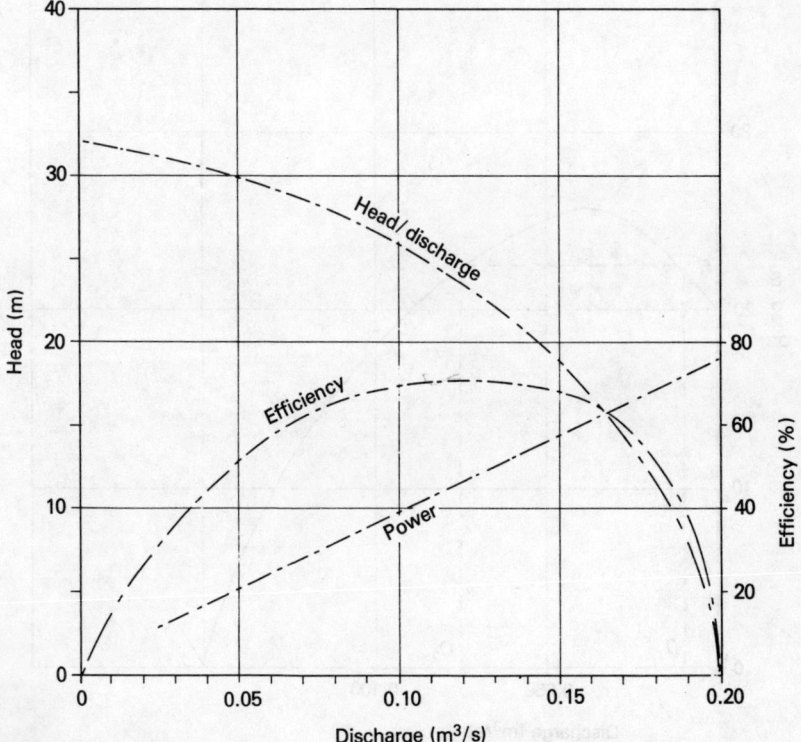

Fig. 10.10 Characteristic curves for a centrifugal pump

Manufacturers normally list these properties as three curves – the characteristic curves – on a single graph (Fig. 10.10), and the best operating point of the pump can be identified as the head, discharge and power values at the maximum efficiency.

It is always worth studying characteristic curves with care and asking the following questions:

1. Is the head-discharge curve reasonably steep? If it is not (see Fig. 10.11), then two discharge rates are possible for the same output head. This can lead to uncontrolled variation of the discharge (the pump is said to 'hunt') and means that a guaranteed flow rate cannot be given. If a number of identical pump units discharge into a single pumping main, such hunting can create very serious problems with surge pressures being generated.

2. Does the efficiency fall off rapidly for quite small variations in the discharge rate?

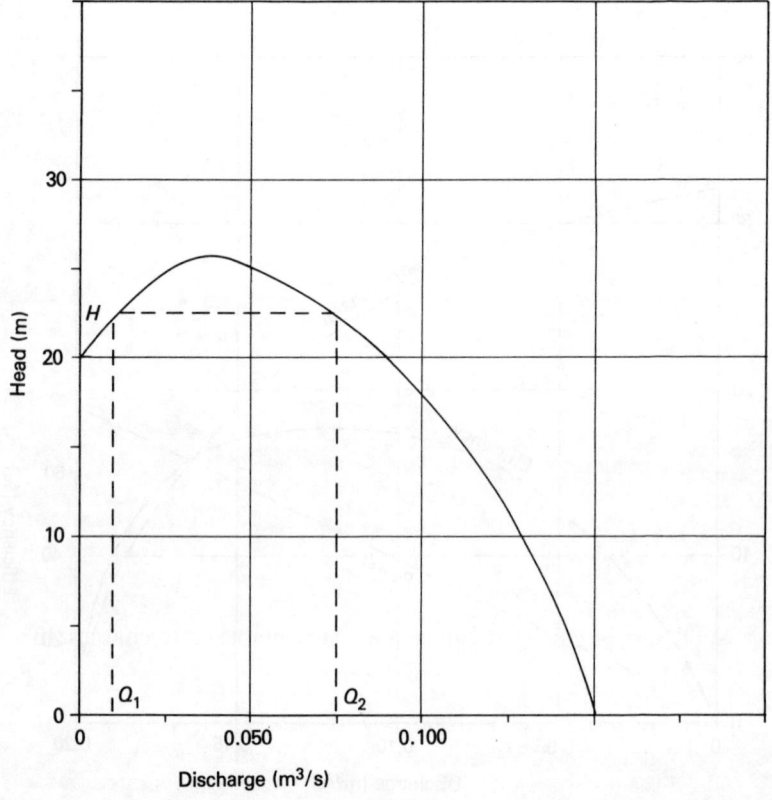

Fig. 10.11 Centrifugal pump with unstable head/discharge characteristics

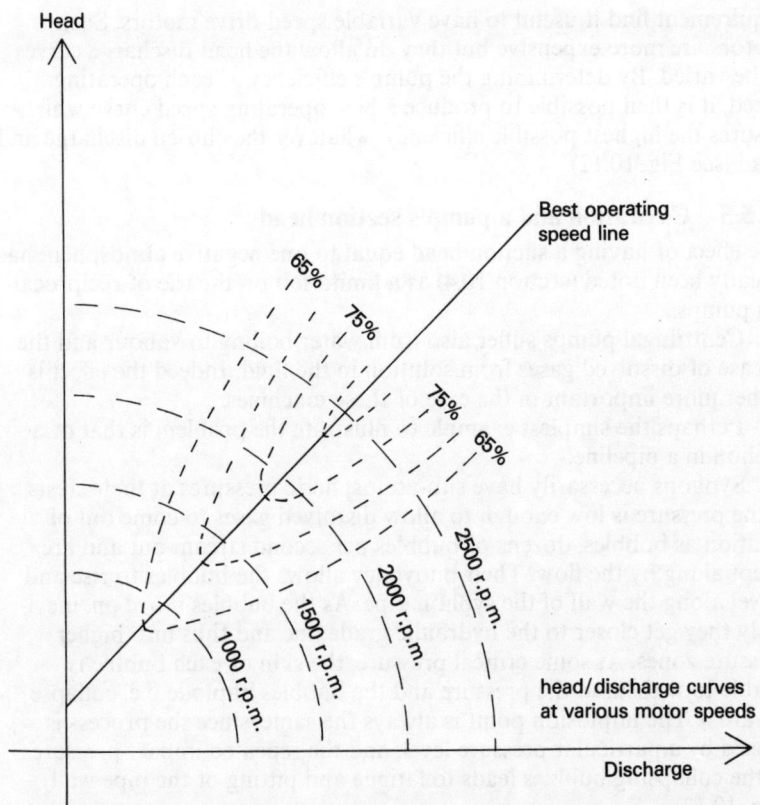

Fig. 10.12 Head, discharge and efficiency when a multi-speed drive motor is used

In many operations, it will be necessary to vary the flow rate, and it is desirable to be able to do this for as little loss of efficiency as possible. On a simple site dewatering, where the pumping costs are only an insignificant part of much larger costs for other engineering operations, this efficiency loss is probably of no real consequence. But those operators (water supply authorities, drainage boards, managers of large chemical plants) who have to pump day in and day out and whose overall expenditure is seriously influenced by pumping costs, find that even a quite small loss of efficiency can create unacceptable extra charges. In such cases it is better not to change the flow rate by varying the valve opening on a discharge pipe; instead it is normal to have a number of pumping units available. Each is run at the maximum efficiency and the discharge rate is altered by starting or stopping individual pumps.

Whilst the great majority of centrifugal pumps are operated by electrical motors of fixed speed, the organisations with a major pumping requirement find it useful to have variable speed-drive motors. Such motors are more expensive but they do allow the head-discharge curves to be varied. By determining the pump's efficiency at each operating speed, it is then possible to produce a best operating speed curve which ensures the highest possible efficiency, whatever the chosen discharge and head (see Fig. 10.12).

10.5.5 Cavitation and a pump's section head

The effect of having a suction head equal to one negative atmosphere has already been noted (section 10.4) as a limitation on the use of reciprocating pumps.

Centrifugal pumps suffer also from water boiling to vapour and the release of dissolved gases from solution in the fluid. Indeed the effect is rather more important in the case of these machines.

Perhaps the simplest example to illustrate the problem is that of a syphon in a pipeline.

Syphons necessarily have sub-atmospheric pressures at their crests. If the pressure is low enough to allow dissolved gases to come out of solution as bubbles, dozens of bubbles per second stream out and are swept along by the flow. Their buoyancy allows the bubbles to rise and travel along the wall of the syphon pipe. As the bubbles move on, inevitably they get closer to the hydraulic grade line and thus into higher pressure zones. At some critical pressure, the skin of each bubble is unable to withstand the pressure and the bubbles implode (i.e. collapse inwards). The implosion point is always the same, since the process is caused by a particular pressure level, and the repeated impact pressures of the collapsing bubbles leads to fatigue and pitting of the pipe wall (Fig. 10.13).

A similar effect takes place in centrifugal pumps. The lowest pressure occurs at the pump eye. As the fluid passes through the impellers, its pressure head rises and a point is reached where bubble implosion occurs. The consequence is that the thin impeller blades are battered and vibrated and can suffer serious damage.

The separation of gas from a liquid is termed **cavitation** and is always accompanied by noise, vibration and a pulsing discharge. Thus it should be easily identified and the pump immediately switched off.

The damage caused by cavitation can be reduced by using materials for impeller blades that are more resistant to abrasion. Cast aluminium bronze, for example, is only 1/30th as susceptible to cavitation pitting as is mild steel.

However, by far the best protection lies in designing the suction arrangements to avoid cavitating negative pressures.

A Bernoulli analysis of the energy at the inlet to the pump and that at its eye produces

(pressure head + velocity head + positional head)$_{inlet}$ =

(pressure head + velocity head + positional head)$_{pump\ eye}$

+ energy losses in suction pipe

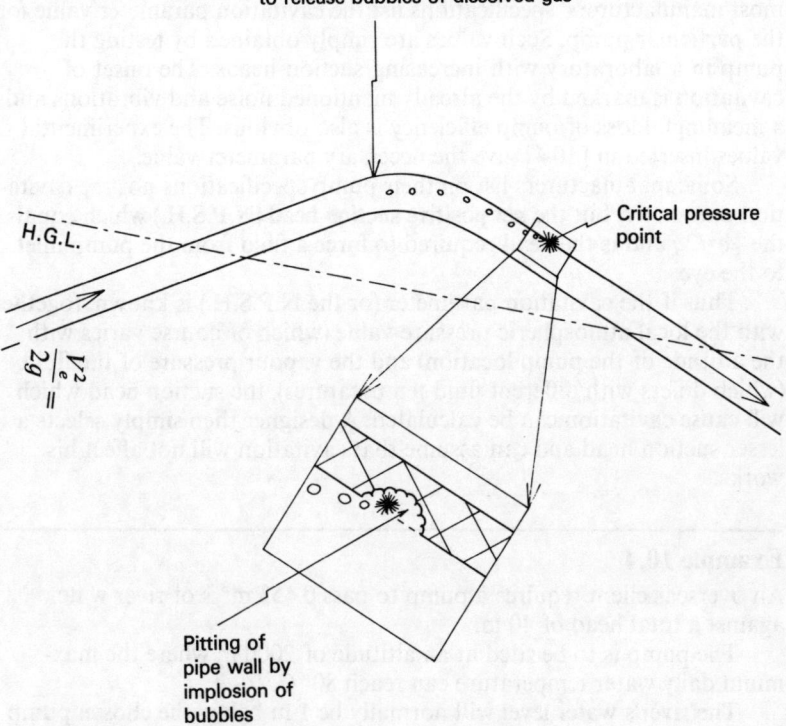

Fluid pressure low enough
to release bubbles of dissolved gas

Critical pressure
point

H.G.L.

$= \dfrac{V^2e}{2g}$

Pitting of
pipe wall by
implosion of
bubbles

Fig. 10.13 Cavitation in a pipe system

and, making all pressures absolute values, thus gives

$$\frac{P_{(atmospheric)}}{\rho g} + 0 + 0 = \left(\frac{P_{(eye)}}{\rho g} + \frac{V^2_{(eye)}}{2g} + Z_{(eye)}\right) + losses$$

or

$$\frac{V^2e}{2g} = \frac{P_{atmospheric} - P_{eye}}{\rho g} - (Z_{(eye) + losses)}$$

For cavitation to take place, the pressure at the pump's eye must be at least as low as the vapour pressure ($P_{(vap)}$) of the fluid (i.e. the pressure at which a liquid changes to a gas)

∴   for cavitation,

$$\frac{Ve^2}{2g} = \frac{(P_{atmos} - P_{vap})}{\rho g} - (Z_{eye + pipe\,flow\,losses})$$

or $\dfrac{Ve^2}{2gH} = \dfrac{1}{H}\left[\dfrac{(P_{atmos} - P_{vap})}{\rho g} - (Z_{eye + pipe\,flow\,losses})\right]$ [10.4]

where H = the total head delivered by the pump.

The term $Ve^2/2gH$ is generally called the **cavitation parameter** (σ) and most manufacturers' specifications list the cavitation parameter value for the particular pump. Such values are simply obtained by testing the pump in a laboratory with increasing suction heads. The onset of cavitation is marked by the already mentioned noise and vibrations and a meaningful loss of pump efficiency is also obvious. The experimental values inserted in [10.4] give the necessary parameter value.

Some manufacturers list on their pump specifications not the cavitation parameter, but the **net positive suction head** (N.P.S.H.) which equals the $Ve^2/2g$ and is the head required to force a fluid from the pump inlet to the eye.

Thus if the cavitation parameter (or the N.P.S.H.) is known, together with the local atmospheric pressure value (which of course varies with the altitude of the pump location) and the vapour pressure of the fluid (which differs with different fluid temperatures), the suction head which will cause cavitation can be calculated. A designer then simply selects a lesser suction head and can assume that cavitation will not affect his works.

Example 10.4

An overseas client requires a pump to pass 0.453 m³/s of river water against a total head of 40 m.

The pump is to be sited at an altitude of 2000 m where the maximum daily water temperature can reach 80 °C.

The river's water level will normally be 1 m below the chosen pump site, though in extreme droughts the level can be 3 m below the pump.

A suction pipe of 0.600 m internal diameter (described by H.R.S. tables for $k = 0.06$ mm) has been chosen.

If the pump's cavitation parameter is 0.05, will the pump be free from cavitation effects?

(Take the local absolute atmospheric pressure as 8.107 m of water, and the vapour pressure of water at 80 °C as 4.826 m of water.)

Solution

From the H.R.S. chart for $k = 0.06$ mm, the hydraulic gradient for a flow of 0.453 m³/s in a 600 mm diameter pipe is –

0.32 m for each 100 m of pipe

(a) **Normally** the river level is 1.0 m below the pump, and in this case,

the suction head = the gravity lift +
head losses in 1 m of suction pipe
= 1.0 m + 0.0032 m
= 1.0032 m

From the values of atmospheric and vapour pressure given

$$\frac{\dfrac{(P_{atmos} - P_{vap})}{\rho g} - (Z + \text{head losses})}{H} = \frac{(8.107 - 4.826) - (1.0032)}{40}$$

$$= \frac{2.2778}{40} = 0.0569$$

As this value is greater than the cavitation parameter, the pump will be free of cavitation effects in the normal operating conditions.

(b) In **extreme** droughts, the river water level will be 3.0 m below the pump.
Thus, for such periods,

$$\frac{\dfrac{P_{atmos} - P_{vap}}{\rho g} - (Z + \text{head losses}) \text{ in 3.0 m of suction pipe}}{H}$$

$$= \frac{(8.107 - 4.826) - (3.0 + 0.0096)}{40}$$

$= 0.00679$, a value less than that of the cavitation parameter.

Thus in extreme drought conditions, when the river level drops, the pump would suffer cavitation.

The solution to this problem is either to lower the pump setting, or to halt pumping in drought conditions, or to keep the river level in drought conditions higher by building a low (2.0 m) dam across the river.

The first option would require a submersible pump, since the normal water level precludes lowering the pump more than 1 m. The second option is unlikely to be acceptable since water supplies are all the more essential in a drought. The third option may well be possible if the construction costs are reasonable.

Example 10.5

If the maximum river water temperature at the pumping station described in Example 10.4 had been 40 °C, would the pump have been affected by cavitation? Assume all other factors remain the same (the vapour pressure head of water at 40 °C is 0.76 m).

Solution
In this case

$$\frac{\dfrac{P_{(atmos)} - P_{(vap)}}{\rho g} - (Z + \text{head losses})}{\text{pumping head}} = \frac{8.107 - 0.76 - 3.0096}{40}$$

for the drought level in the river.

166

This equals 0.108 and is greater than the cavitation parameters. Similarly, for the higher river level, the pump would not be affected by cavitation.

The maximum temperature of the water to be pumped is thus more important than it would first appear.

Example 10.6

The proposed suction arrangements for a river abstraction pump are shown on Fig. 10.14.

Comment critically on the proposals and suggest at least one better solution.

Solution

The suction pipeline – as designed – is not only long (high friction losses) but has a right angle bend (large minor losses) where air blockage is likely to occur.

This is guaranteed to produce quite large negative pressure on the suction side of the pump and is undesirable.

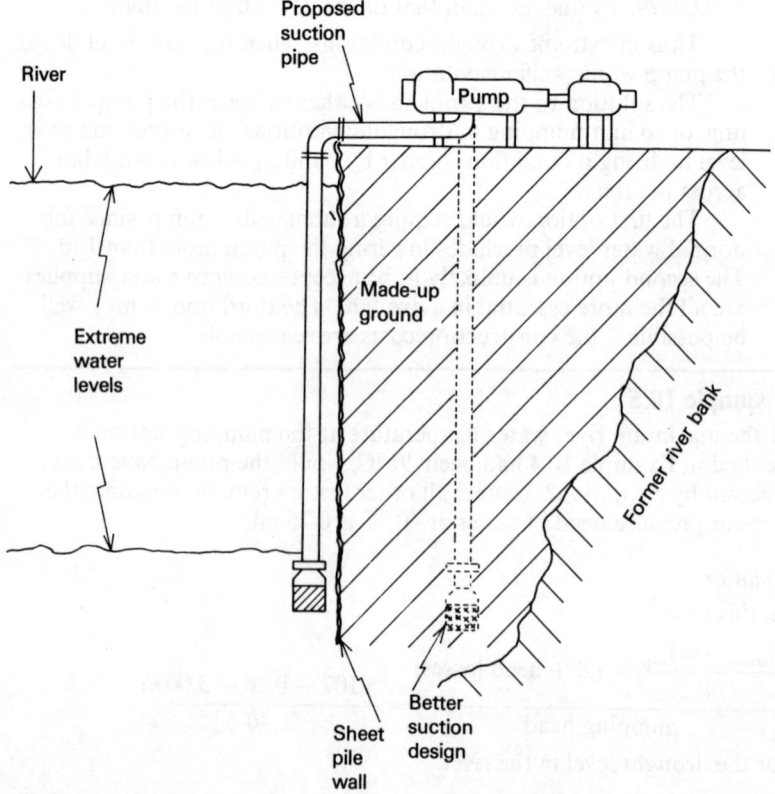

Fig. 10.14 Suction arrangements at proposed abstraction works

A better arrangement (see Fig. 10.14) could be:

(a) To reduce the suction pipe length and remove the right angled bend. This would improve the suction head losses and give a lower possibility of cavitation effects.
(b) Even better, if it is practically and economically possible, is to avoid all negative pressures on the suction side of the pump. This can be done by excavating a dry well (below the lowest water level in the river) and running the inlet pipe through a waterproof gland to the pump. In this way, a positive head will always exist on the suction pipe and fears of cavitation can be ignored. An additional advantage, in countries with environmental legislation, is that the pumping station is buried underground.

10.5.6 Augmenting discharge and pumping head
If a pump's output is inadequate (either in flow rate or pumping head) it can, of course, be replaced by a more powerful unit.

In many cases, this solution is not particularly useful since the original pump may only be inadequate for relatively short periods. An obvious example is that of a holiday town's water supply system where demand is fairly low for most of the year, and only increases significantly for a few weeks in the summer tourist season.

If such a situation exists it is often worth installing two or more pumps on the same pumping main. Such pumps can be **in parallel** or **in series** (Fig. 10.15).

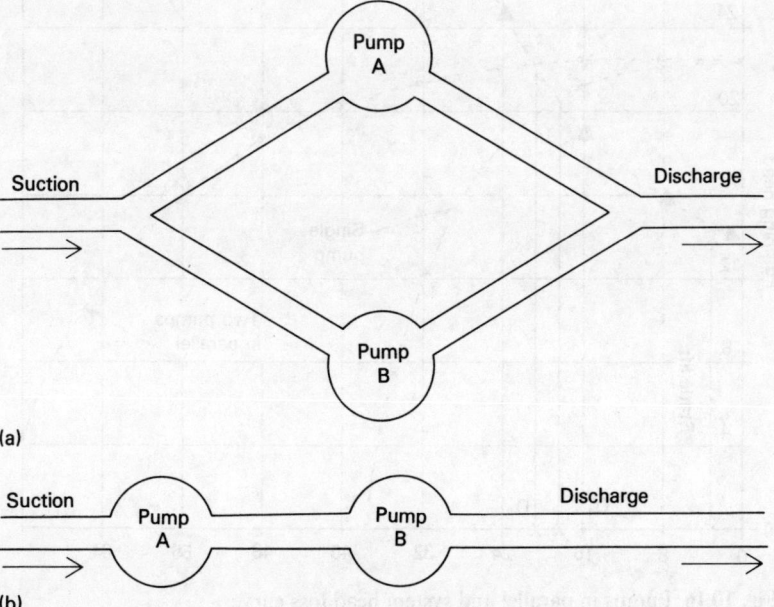

Fig. 10.15 Pumps in parallel and in series. (a) Plan of parallel pump installation   (b) Plan of series pump installation

168

With parallel pumps, each unit's output is taken into the same discharge pipe. Thus a parallel installation is a method of increasing the flow rate and it does not affect, in any way, the pumping head produced.

On the other hand, pumps in series are arranged to pass the output from one unit into the suction side of the other pump. Thus exactly the same discharge as for a single pump results, and all that is altered is the pumping head.

The operating curves for pumps in parallel or in series are easily obtained from the manufacturer's characteristic curves for the single pump. With two, or more (say n) pumps set up in parallel, the discharge produced for any particular head is simply twice, or n times, that for the single pump. Similarly for a series installation, the head at any specified discharge rate is twice, or n times, that of the single unit (see Fig. 10.16).

Thus it might be though that two parallel pumps, **in a pipeline**, will produce twice the original discharge.

This, however, overlooks the basic point that the frictional head losses, for flow in a pipeline, increase with the amount of the discharge.

An example will make this clearer.

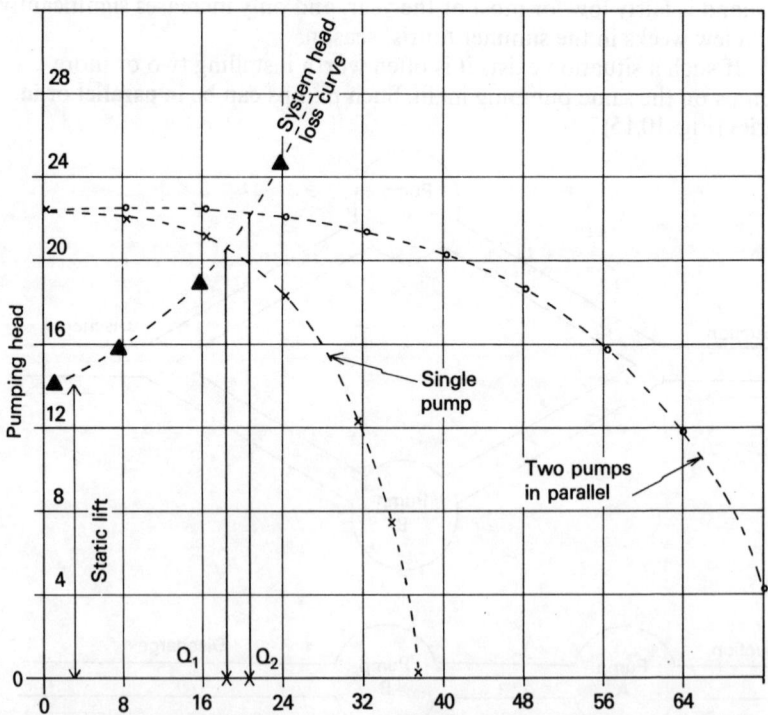

Fig. 10.16 Pumps in parallel and system head loss curve

Example 10.7

A particular pump, when tested by the manufacturer, produced (see Fig. 10.16)

Head (m)	22.5	22.2	21.6	19.5	14.1	0
Discharge $(m^3/s \times 10^{-2})$	0	7.5	15	22.5	30	37.5

The pump is to be installed in a pipeline 75 m long (71 m of delivery pipe and 4 m of suction pipe) and is to lift water 10 m above the pump. The pump itself is 4 m above the source water.

The delivery pipe is 0.250 m in diameter and the suction pipe is somewhat larger, 0.350 m in diameter. All pipes are described by a D'Arcy f of 0.008.

What percentage increase in flow would two pumps, installed in parallel, produce?

Solution

1. The first necessity is to draw the characteristic curve for the single pump and then multiply its discharge valves, by two, to obtain the head/discharge relationships for two parallel units (Fig. 10.16).

2. It is then imperative to determine what head losses will occur, for particular discharges, in the specified pipeline:

   (a) At zero discharge, the static lift of the system is the sum of the discharge and suction gravity lifts,

   i.e. $\underline{10\,m + 4\,m = 14\,m}$

   (b) At any other chosen discharge (say 0.165 m³/s) the frictional head loss is

   $$H = \frac{64 \times 0.008 \times 71 \times 0.165^2}{2 \times 9.81 \times 0.25^5 \times \pi^2} + \frac{64 \times 0.008 \times 4 \times 0.165^2}{2 \times 9.81 \times 0.35^5 \times \pi^2}$$

   $$= \frac{0.990}{0.189} + \frac{0.056}{1.018}$$

   $$= 5.293$$

   ∴ total pumping head $= 5.293 + 14 = \underline{19.293 \text{ m}}$

   (c) At any other discharge (say 0.240 m³/s) similarly, the frictional head loss is

   $H = 11.195$ m

   and the total pumping head $= \underline{25.195 \text{ m}}$

   (d) Joining these three calculated head valves, with a smooth curve, produces the **system head loss curve**, i.e. the line marking out the variation in total head required, for different discharges, in this particular pipe system.

Where this cuts the H/Q curves, for the single pump and the two parallel pumps, the flow rates that will be produced can be read off as:

for single pump – 0.184 m³/s
for parallel pumps – 0.280 m³/s

Thus the increase in flow rate is 52.17 per cent

a value much less than that initially expected.

A point always worth checking with a parallel pump operation is how the suction head has been increased. In many cases, this extra head will make the parallel units subject to cavitation (see section 10.5.5), whereas the single pump itself was safe from this effect.

Example 10.8

If the parallel pump system (Example 10.7) is to be installed at an altitude of 2000 m, in a location where water temperatures often attain 40 °C, will cavitation effects occur? (assume that the cavitation parameter (σ) is 0.15).

Solution

Using the atmospheric and vapour pressure data specified in Example 10.5, and having calculated the suction head losses for 0.184 m³/s and 0.280 m³/s (4.068 m and 4.158 m respectively), for a single pump,

$$= \frac{8.107 - 0.76 - 4.068}{20.48}$$

$$= 0.160$$

and for the parallel pumps

$$= \frac{8.107 - 0.76 - 4.158}{22.28}$$

$$= 0.143$$

Thus the single pump would not be subject to cavitation, but the parallel pumps would be affected.

A very similar set of calculations can be carried out for a proposed in series pump installation.

10.5.7 The design of a pumping main

The basic design concept for a gravity main was that all the available head had to be utilised in overcoming friction and minor losses (Ch. 7). This ensured that the smallest, and thus cheapest, pipe diameter was chosen to pass the required flow.

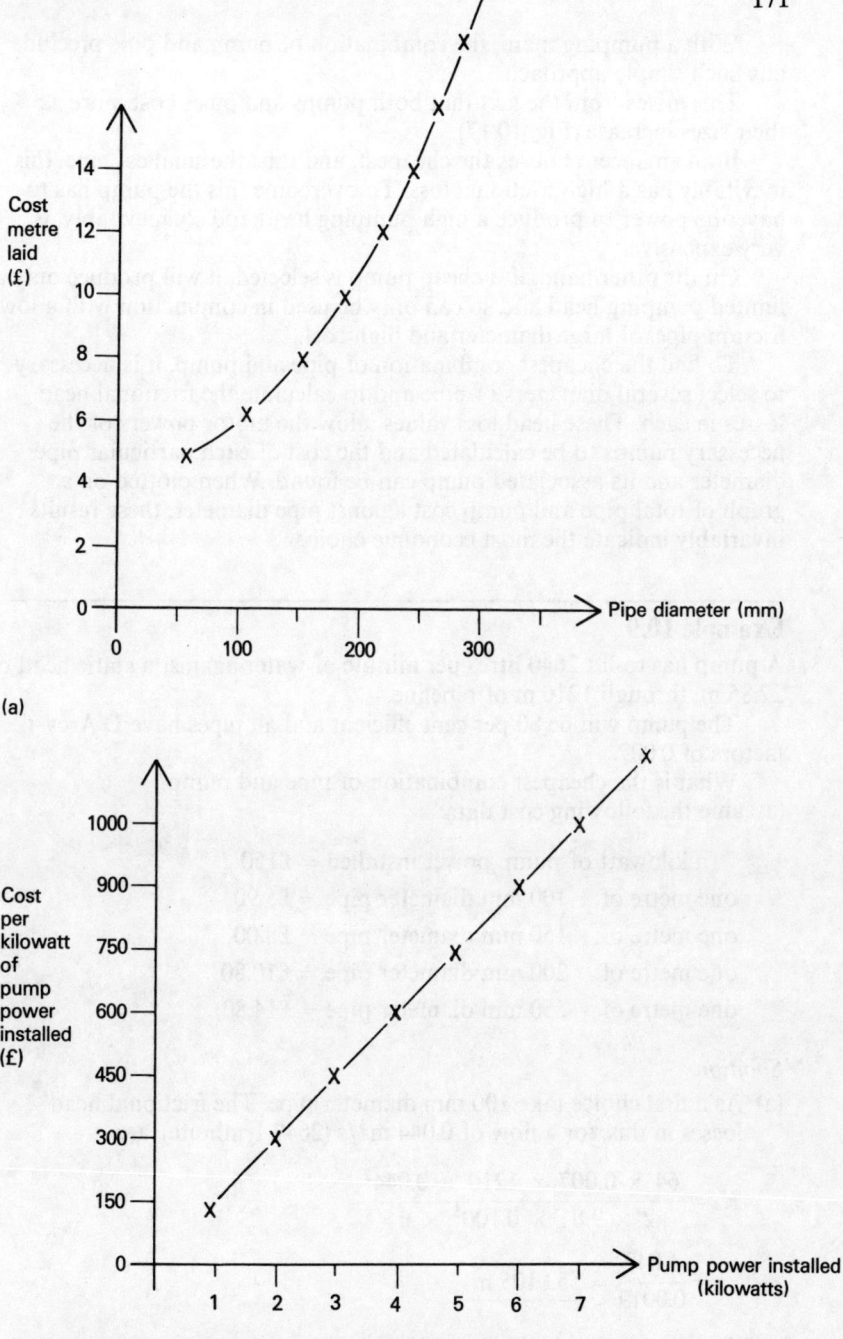

(a)

(b)

Fig. 10.17 Factors affecting the economic choice of a pumping main. (a) Variation of pipe costs with diameter (in 1978) (b) Variation of pump cost with motor power (in 1978)

With a pumping main, the combination of pump and pipe preclude any such simple approach.

This arises from the fact that both pumps and pipes cost more as their sizes increase (Fig. 10.17).

If an engineer chooses the cheapest, and thus the smallest pipe, this inevitably has a high frictional loss. To overcome this the pump has to have the power to produce a high pumping head and so, inevitably, is very expensive.

On the other hand, if a cheap pump is selected, it will produce only a limited pumping head and so can only be used in conjunction with a low friction pipe, of large diameter and high cost.

To find the cheapest combination of pipe and pump, it is necessary to select several diameters of pipe and to calculate the frictional head losses in each. These head loss values allow the motor powers of the necessary pumps to be calculated and the cost of each particular pipe diameter and its associated pump can be found. When plotted on a graph of total pipe and pump cost against pipe diameter, these results invariably indicate the most economic choice.

Example 10.9

A pump has to lift 2640 litres per minute of water against a static head of 22.85 m, through 1210 m of pipeline.

The pump will be 80 per cent efficient and all pipes have D'Arcy f factors of 0.007.

What is the cheapest combination of pipe and pump?
(assume the following cost data:–

> a kilowatt of pump power installed = £150
> one metre of: 100 mm diameter pipe = £5.90
> one metre of: 150 mm diameter pipe = £8.00
> one metre of: 200 mm diameter pipe = £10.80
> one metre of: 250 mm diameter pipe = £14.80)

Solution

(a) As a first choice take 100 mm diameter pipe. The frictional head losses in this, for a flow of 0.044 m³/s (2640 1/minute), are

$$H = \frac{64 \times 0.007 \times 1210 \times 0.044^2}{2 \times 9.81 \times 0.100^5 \times \pi^2}$$

$$= \frac{1.049}{0.0019} = \underline{552.105 \text{ m}}$$

thus the total pumping head = 552.105 + 22.85 m

$$= \underline{574.955 \text{ m}}$$

The power needed by the pump to pass the design flow is

$$P = (\rho \times g \times Q \times H)(\text{efficiency}) \qquad [10.2]$$

$$= (1000 \times 9.81 \times 0.044 \times 574.955)\left(\frac{100}{80}\right) \text{watts}$$

$$= \underline{310.22 \text{ kilowatts}}$$

Note that the efficiency is included as

$$\frac{100}{80}, \quad \text{not as} \quad \frac{80}{100}$$

This reflects the fact that the pump is 20 per cent inefficient and so has to receive 20 per cent more motor power than the hydraulic head it produces.

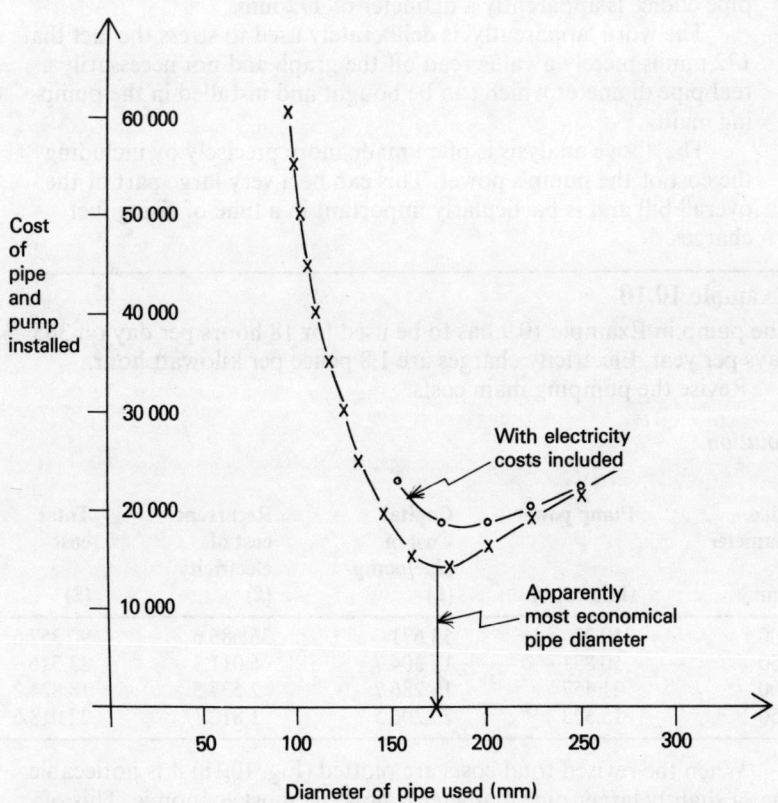

Fig. 10.18 Determining the most economical combination of pump and pipe diameter

The costs of this particular pipe/pump combination are

pipe: $1210 \times £5.90 = £\ 7139$

pump: $310.22 \times £150 = £46\,532$

$$\text{Total} = \underline{\underline{£53,671}}$$

(b) The same procedure is followed for at least three other pipe diameters.

Pipe diameter (mm)	Head lost (m)	Total head (m)	Pump power (kilowatts)	Cost of pipe & pump (£)
250	5.550	28.40	15.323	20 206.5
200	16.920	39.77	21.457	16 286.69
150	71.360	94.21	50.831	17 304.69

From the graph of the results in Fig. 10.18 the most economical pipe choice is apparently a diameter of **172 mm**.

The word 'apparently' is deliberately used to stress the fact that 172 mm is merely a value read off the graph and not necessarily a real pipe diameter which can be bought and installed in the pumping main.

The above analysis is often made more precisely by including the cost of the pump's power. This can be a very large part of the overall bill and is particularly important in a time of rising fuel charges.

Example 10.10

The pump in Example 10.9 has to be used for 18 hours per day on 365 days per year. Electricity charges are 1.8 pence per kilowatt hour.

Revise the pumping main costs.

Solution

Pipe diameter (mm)	Pump power (kilowatts)	Capital Cost of pipe/pump (£)	Recurrent cost of electricity (£)	Total cost (£)
100	310.22	53 671	36 686.6	90 357.6
150	50.831	17 304.7	6 011.3	23 316
200	21.457	16 286.7	2 537.5	18 824.2
250	15.323	20 206.5	1 812.1	22 018.6

When the revised total costs are plotted (Fig. 10.18) it is noticeable that a slightly larger pipe diameter is now the most economic. This of course is as expected, since head losses and pump power, and so electricity demand, decrease as the pipe diameter is made greater.

Once the choice of pump and pipeline has been made, the remainder of the design consists of noting the points made earlier. Amongst the more important of these are –

(a) the head losses in the suction pipe and the risk of cavitation;
(b) the need to protect the fragile impellers of the pump from any surge pressures, should the power supply fail abruptly;
(c) the lower power requirements at zero discharge. To avoid the need for large electrical currents, pumps are normally started against a closed discharge valve, installed just upstream of the pump.

Section 10.5 – Self assessment questions

1. A scroll case is essential for the operation of a simple centrifugal pump. Why is this?
2. What is meant by 'priming' a centrifugal pump? Is this an essential operation?
3. Is the discharge produced by a centrifugal pump related to the pumping head it develops?
4. Define the term 'pumping head'. Does it include both the delivery and the suction head?
5. Can a positive head be generated in the suction pipe to a centrifugal pump? Is there any advantage in achieving this?
6. Draw the characteristic curves for a simple centrifugal pump. What should be the chosen operating point?
7. If liquid water converts to water vapour at a water pressure of -10.3 m (at sea-level), why should cavitation effects appear at smaller negative heads?
8. Define cavitation parameter and net positive suction head.
9. What factors influence whether or not cavitation will occur?
10. Should a suction pipe to a centrifugal pump be as short, straight and frictionless as possible? Explain why.
11. From any suitable book, draw a 'dry well' pump installation, which ensures that a positive head occurs across the pump's suction pipe.
12. From the single pump characteristic curve (Fig. 10.16) draw the operating curve produced by three identical pumps installed in series. Superimpose the system head loss curve using the data of Question 10.7 and determine the flow rate and pumping head that can be obtained.
13. Clearly distinguish between the design methods for a gravity pipe and a pumping main. Why is the pumping main case more complicated?

10.6 Axial and mixed flow pumps

Although the centrifugal pump is a flexible and useful civil engineering tool, it does suffer from one inbuilt limitation.

This can be appreciated if the flow paths of the fluid from the eye of the pump along the impeller blades are considered. (Fig. 10.19).

The fact that the fluid particles have to turn through 90° imposes a real limit on the volume that such a pump can handle. This situation will be recognised as essentially similar to the case of a 90° bend in a pipe

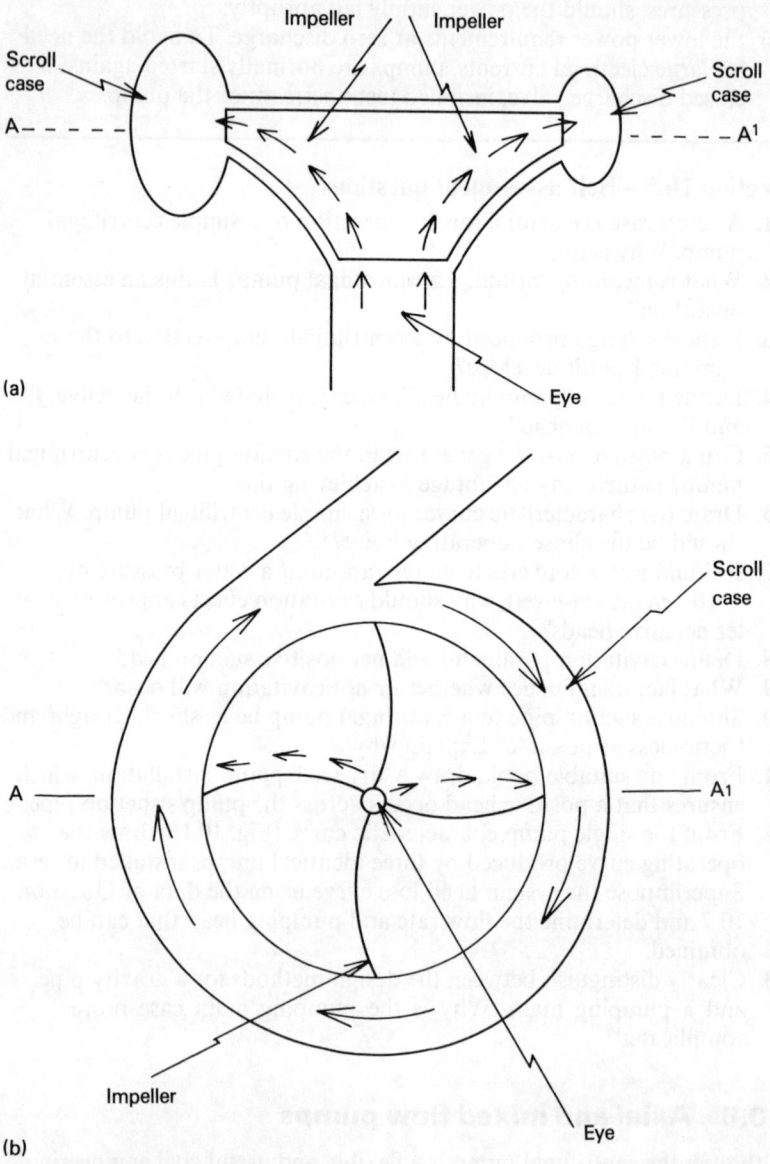

(a)

(b)

Fig. 10.19 Water flow paths through centrifugal pump

(Fig. 7.4d), where the inertia of the fluid leaves only a part of the pipe's area to carry flow.

In situations where large volumes of water have to be moved in short time – for example, in flood relief operations – a centrifugal pump is thus inappropriate and use is made of **axial flow** machines.

These (Fig. 10.20) avoid the limitation of the centrifugal devices by diverting the flow paths of the water as little as possible. However, since the advantage of centrifugal force is lacking in such pumps, the larger discharges are produced only with moderate pumping heads (up to 15 m).

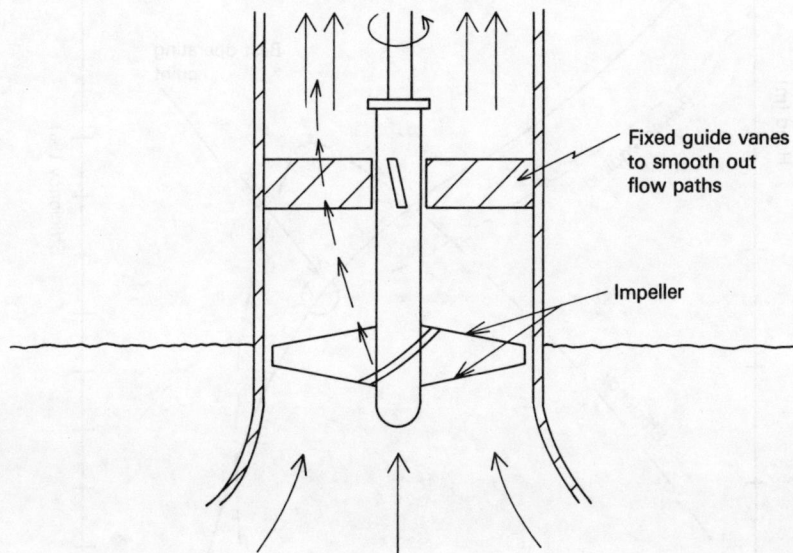

Fixed guide vanes to smooth out flow paths

Impeller

Fig. 10.20 Axial flow – low lift pump

The most widespread use of axial flow pumps is thus with those public bodies responsible for land drainage in low-lying areas.

As a consequence of their quite different basis of operation, the characteristic curve of an axial flow machine is dissimilar from that of a centrifugal pump (Fig. 10.21), and, in particular, the power requirement at zero discharge should be noted. Because the power demand is so high at very low flows, a risk of overloading and burning out the electric motor exists. To avoid this, it is normal to operate axial flow machines without any inlet valve.

Cavitation is as serious a risk as in centrifugal pumps and generally axial flow machines are installed with the impellers below the water's surface.

The advantage of the axial flow pumps of high discharge rate is largely negated, for most normal uses, by the very low pumping head,

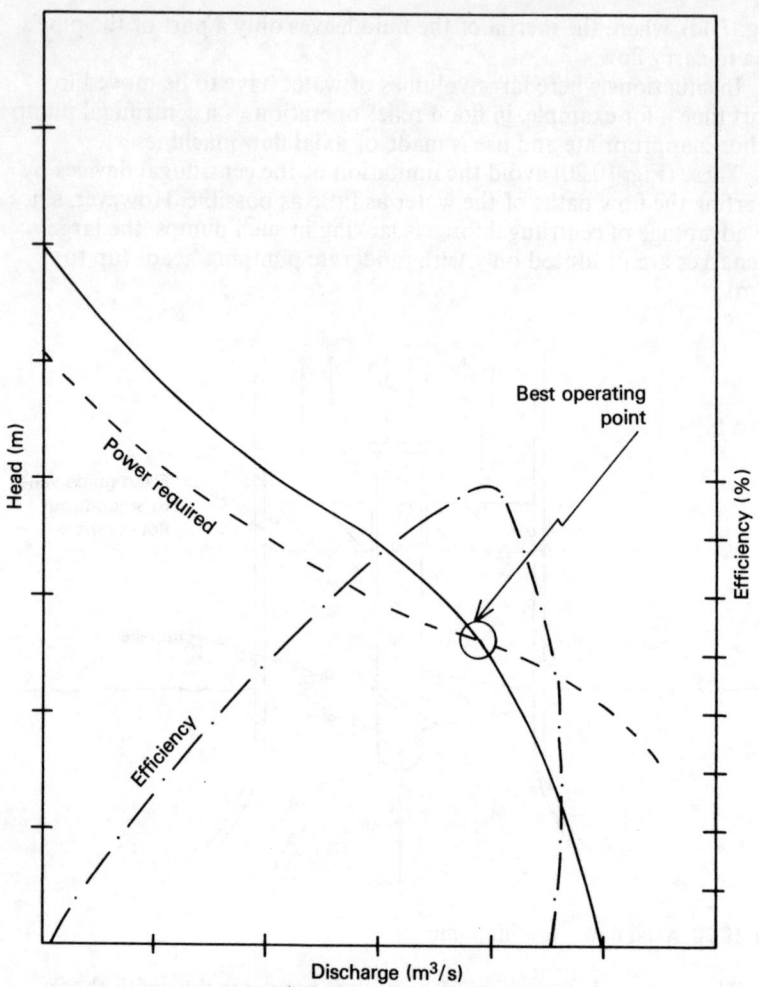

Fig. 10.21 Characteristic curve of axial flow pump

which obviously must result from a pumping action which consists essentially of lifting slices of water by curved propeller-like impellers. Because of this, a number of manufacturers have built **mixed flow** pumps, which have impellers rather like those of centrifugal machines. The broader impellers installed in these devices reduce the change in direction experienced by the water particles and still give some measure of centrifugal force, with its advantages of accelerating the water and increasing its velocity head. The properties of such pumps, and their characteristic curves, lie therefore between those of the two major groups of rotodynamic pumps.

10.7 Hydroelectric turbines

Whilst centrifugal and axial flow pumps utilise motor power to drive impellers, and so generate hydraulic head, turbines perform precisely the reverse action and remove hydraulic energy which is then converted to electrical power.

In many countries, particularly in recent years with higher costs of petroleum products, hydro-electricity has become a major part of the energy supply and one that has given employment to large numbers of civil engineers.

The two major classes of turbines are the impulse and reaction machines.

Impulse turbines are the simplest and cheapest and are simply a logical development of the older water wheel. The water's energy is converted into velocity head (by a suitable nozzle on the end of the water supply pipeline) and the resulting high-speed jet is allowed to strike curved buckets on the periphery of a wheel. The rotation of the wheel spins the magnetic core of the associated generator and so creates the desired electrical current (Fig. 10.22).

The buckets are usually made as twin units (Fig. 10.22b) arranged on either side of a jet splitter which allows equal flow to strike each half of

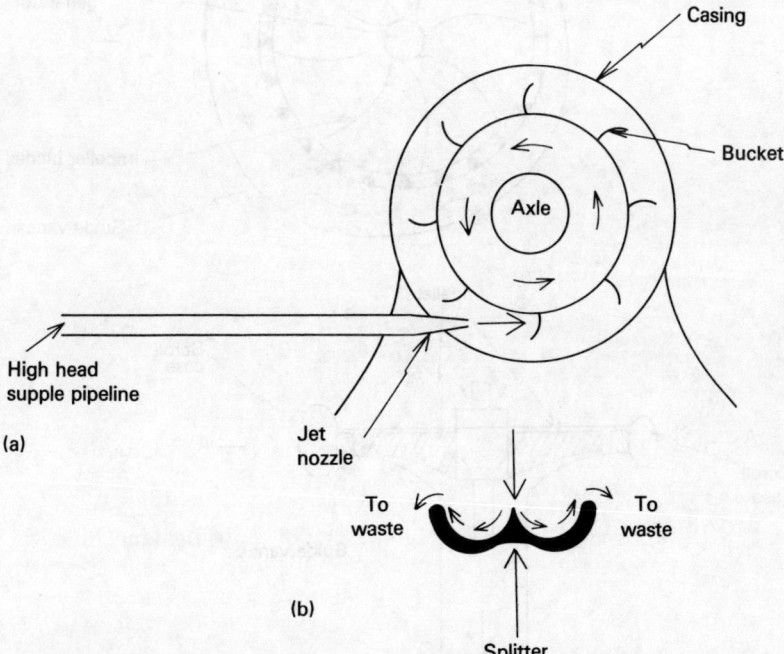

Fig. 10.22 Single jet impulse turbine of Pelton wheel type. (a) General details (b) Bucket details

the bucket. After the energy of the water has been converted into rotation of the wheel, the waste water is allowed to run off the buckets for disposal.

The casing around the impulse turbine is simply there to prevent water spraying all over the turbine house.

The most widespread modern variety of impulse machine is the Pelton wheel, though older designs (resembling the common lawn sprinklers) still exist.

To operate effectively, impulse turbines need a very high input hydraulic energy and, as a result, they tend to be installed only in mountainous regions, where heads of at least 150 m are available.

Most reaction turbines resemble the already described centrifugal pumps and are much more expensive and complex than the Pelton wheel (for which no pump equivalent exists).

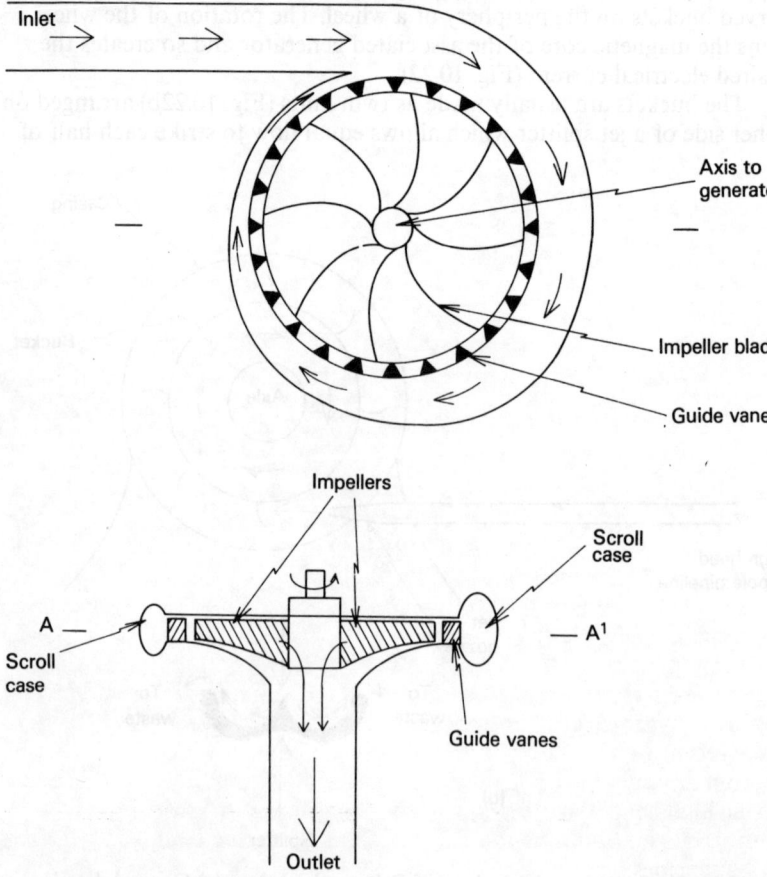

Fig. 10.23 Francis type reaction turbine

In general terms, they operate by allowing water from the scroll case to enter the impellers via adjustable guide vanes. These vanes exist both to ensure that turbulence does not occur, which would, of course, degrade some of the water's energy, but also to convert some of the fluid's pressure energy to velocity head. The water then flows through the impeller blades and a torque (which rotates the turbine axis and thus the generator element) is caused by the deviation in the direction of the water's flow and the change in pressure and velocity head.

To ensure that the water entering all the guide vanes does so with the same velocity, the scroll case has a decreasing cross-sectional area to allow for the reducing quantity of flow around it (Fig. 10.23).

Because of their much larger dimensions and because the power they raise is valuable, the efficiencies of such turbines are rather higher than those of comparable pumps.

The Francis type reaction turbine can operate successfully from heads as low as 15 m to as high as several hundred metres. As a consequence, it is the commonest type of turbine encountered.

Another, and increasingly important, type is the Kaplan machine. This resembles the axial flow pumps and operates in the same limitations–i.e. large flow rate and low head. With renewed interest in generating power from ocean tides, such machines have been recently installed in several stations around the world.

Few civil engineers ever become involved in the design or construction of hydro-electric turbines and those who do work on hydro-electric schemes are generally involved with the necessary dam, tunnels and surge tank. For this reason, only a tentative outline of turbine types has been given. Many excellent and specialised texts do exist which will provide the additional background which will be needed by any engineer whose employers have a major involvement in turbines.

Sections 10.6 to 10.7 – Self assessment questions

1. Clearly explain – in terms of the distortion of flow lines – the particular advantages of an axial flow pump and how these are obtained. Why can a centrifugal pump never discharge as great a volume as an axial flow pump, of equivalent size?
2. Compare the characteristic curves of a centrifugal and an axial flow pump. Why is it poor policy to install a control valve on an axial flow pump?
3. Clearly distinguish between impulse and reaction turbines and state where each is best employed.
4. Why is hydro-electric power generation more popular today than in the recent past?
5. From any suitable book obtain details of two hydro-electric schemes (one high head, the other low head). Identify clearly the different turbines used and note the difference in engineering complexity of the two schemes.

Chapter 11

Flow in open channels

11.1 Introduction

Whilst less common than pressure pipes, open channels are still widely used and enter into the work of a variety of civil engineering specialists. The soils engineer trying to keep an earth embankment stable, the public health engineer designing a sewage treatment plant and the irrigation planner who has to distribute water over large fields all have a need for open channels.

The basic difference from a pressure pipe is simply that in an open channel the fluid's surface is exposed to atmospheric pressure. Thus the hydraulic grade line coincides with the surface of the fluid, and the cross-sectional area of flow decreases and increases, as the discharge rate varies.

11.2 Normal depth flow

The earliest hydraulic engineers noted that, when water entered an open channel, the depth of flow gradually diminished and then became constant throughout the channel, provided that its geometric cross-section and bed slope did not vary.

The reason for this is made clear if the difference in behaviour of ideal and real fluids is considered.

An ideal fluid, lacking viscosity, and so experiencing no frictional drag against the channel boundaries, is accelerated by gravity into the channel (Fig. 11.2a). The increase in the fluid's velocity is, of course, accompanied by a decline in its depth, since for an open channel the

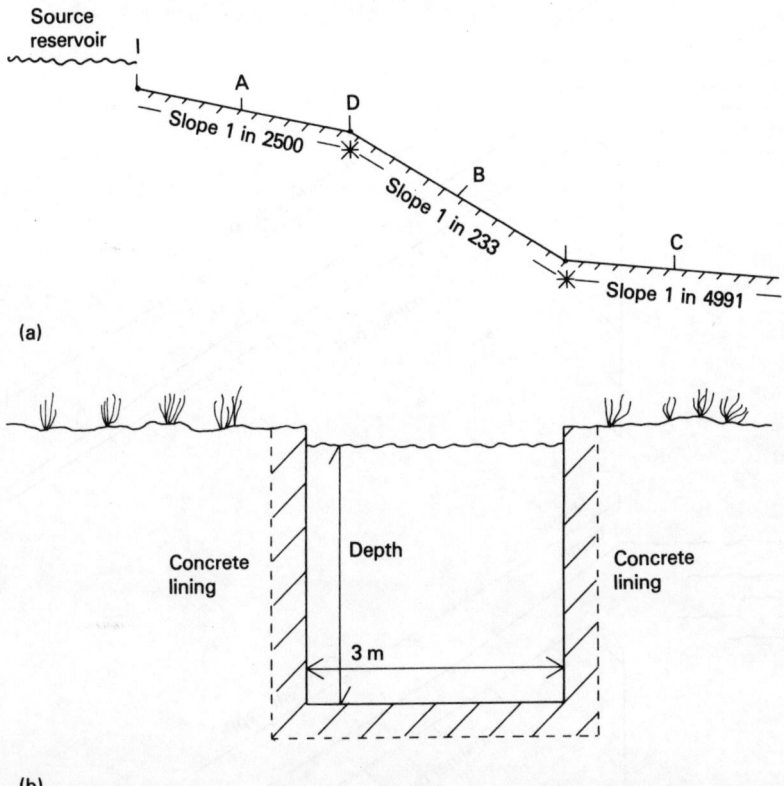

Fig. 11.1 Details of the channel used to illustrate the various open channel flow formulae. (a) Longitudinal section (b) Typical cross-section of channel

equivalent of pressure head in a pipe is the depth of the fluid. As no frictional resistance exists, there is nothing to reduce the initial acceleration, and the fluid continues to speed up till ultimately its depth of flow is infinitely thin.

A real fluid, such as water, initially behaves in the same way, and a length of channel with a decreasing depth of water is noticeable in Fig. 11.2b. However, after some distance of flow, further acceleration is counteracted by the frictional resistance between the water and the channel boundaries. Thus the velocity stabilises, and the depth remains constant, at what is termed the **normal depth**.

Normal depth thus can only occur where a balance exists between acceleration down the channel, and frictional retardation against the flow. In most real life cases this is in the middle reaches of long straight lengths of channel, of reasonably uniform cross-sectional area.

Various early engineers attempted to relate normal depth to the factors which obviously influenced it – the fluid's average velocity, the

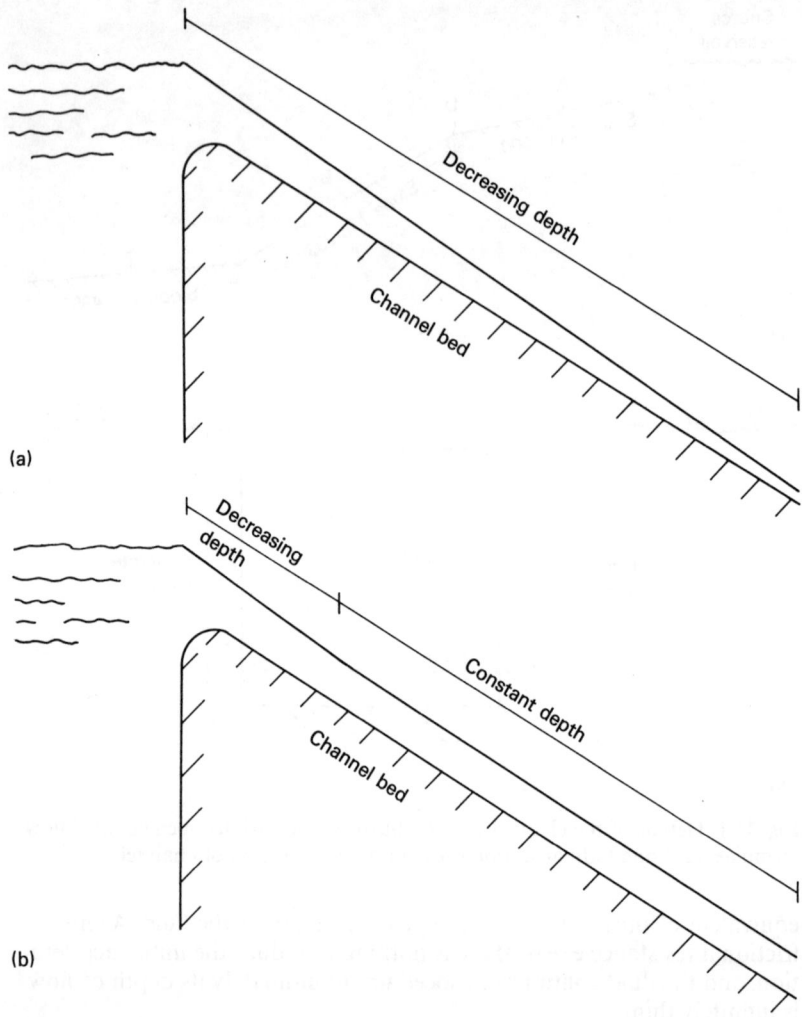

Fig. 11.2 Normal depth of flow in open channels. (a) Ideal fluid flow
(b) Flow of real fluid

geometry of the channel's cross-section, the roughness of the bed
materials and the slope of the channel's bed.

Chezy (in 1775) was the first to succeed, when he produced the
following empirical equation:

$$\bar{V} = C\sqrt{m \times i} \tag{11.1}$$

$$= C\sqrt{\frac{A}{P} \times i}$$

and, for a rectangular channel,

$$\bar{V} = C \sqrt{\frac{w \times d}{w + 2d} \times i}$$

or $\quad Q = A \times C \sqrt{m \times i}$

The meanings of the terms in these equations are shown in Fig. 11.3.

The constant C was included to give a measure of the roughness of the channel's wall materials. However, this constant was soon found to suffer from the limitations already noted for D'Arcy's f factor (Ch. 7), and, to make matters worse, was suitable only for the cross-sectional geometries used by Chezy.

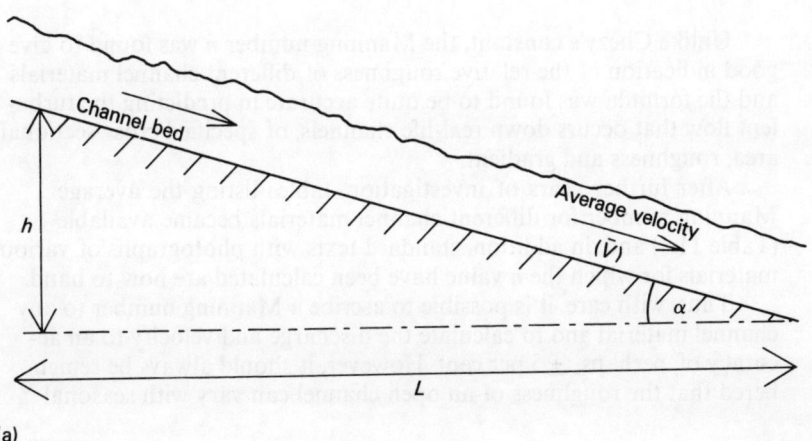

(a)

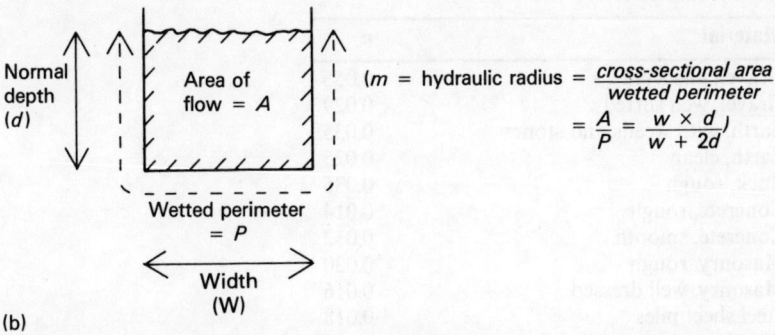

(m = hydraulic radius = $\dfrac{\text{cross-sectional area}}{\text{wetted perimeter}}$

$= \dfrac{A}{P} = \dfrac{w \times d}{w + 2d}$)

(b)

Fig. 11.3 Terms used in the Chezy and Manning equations. (a) Longitudinal section (b) Cross-section

After a further century of investigation, Manning (in 1891) proposed the following variation of Chezy's empirical formula –

$$\bar{V} = \frac{1}{n}(m)^{2/3}(i)^{1/2} \qquad [11.2]$$

$$= \frac{1}{n}\left(\frac{A}{P}\right)^{2/3}(i)^{1/2}$$

(and for a rectangular cross-section)

$$= \frac{1}{n}\left(\frac{w \times d}{w + 2d}\right)^{2/3}(i)^{1/2}$$

or,

$$Q = \frac{A}{n}(m)^{2/3}(i)^{1/2} \qquad [11.3]$$

Unlike Chezy's constant, the Manning number n was found to give a good indication of the relative roughness of different channel materials and the formula was found to be quite accurate in predicting the turbulent flow that occurs down real-life channels, of specified cross-sectional area, roughness and gradient.

After further years of investigation, tables listing the average Manning number for different channel materials became available (Table 11.1) and, in addition, standard texts with photographs of various materials for which the n value have been calculated are now to hand.

Thus, with care, it is possible to ascribe a Manning number to any channel material and to calculate the discharge and velocity to an accuracy of, perhaps, ± 5 per cent. However, it should always be remembered that the roughness of an open channel can vary with seasonal

Table 11.1 Average Manning Numbers (n) for channel materials

Material	n
Gravel, unsorted	0.055
Gravel, well sorted	0.029
Earth, with weeds and stones	0.035
Earth, clean	0.025
Rock, rough	0.035
Concrete, rough	0.014
Concrete, smooth	0.012
Masonry, rough	0.030
Masonry, well dressed	0.016
Steel sheet piles	0.018
Wood, unplaned	0.013
Wood, planed	0.012

factors. Water weeds, for example, grow in summer, choke a channel and increase the *n* factor, and then die out in winter to leave the channel smoother. The choice of the correct *n* value, therefore, requires more experience than might first be thought.

Returning to the channel shown in Fig. 11.1.

Example 11.1

In Fig. 11.1, a flow rate of 15 m^3/s is required along the 3 m wide rectangular concrete channel ($n = 0.012$).

What are the water depths at points –
A, B, and C – which lie at the middle of the three channel gradients?

Solution

For a rectangular section, the Manning equation can be written as

$$Q = \frac{w \times d}{n} \left(\frac{w \times d}{w + 2d}\right)^{2/3} (i)^{1/2}$$

(a) <u>for a bed slope of 1 in 2500</u>

$$15 = \frac{3 \times d}{0.012} \left(\frac{3 \times d}{3 + 2d}\right)^{2/3} \left(\frac{1}{2500}\right)^{1/2}$$

$$= 250d \left(\frac{3d}{3 + 2d}\right)^{2/3} \left(\frac{1}{50}\right)$$

or $\quad 3 = d \left(\frac{3d}{3 + 2d}\right)^{2/3}$

and, by trial and error substitution, of various depth values –

$d =$ the depth at A = 3 m

(b) <u>for a bed slope of 1 in 233</u>

similarly

$$15 = \frac{3 \times d}{0.012} \left(\frac{3d}{3 + 2d}\right)^{2/3} \left(\frac{1}{233}\right)^{1/2}$$

or $\quad 0.916 = d \left(\frac{3d}{3 + 2d}\right)^{2/3}$

and, again by trial and error,

$d =$ the depth of water at B = <u>1.2 m</u>

(c) <u>for a bed slope of 1 in 4991</u>

similarly

$d =$ the water depth at C = <u>4 m</u>

Thus, apart from the nuisance of trial and error substitutions, the equation is simple to use.

Despite this, the Manning equation is not more than an empirical combination of the terms likely to affect the normal depth of flow. In the analysis of pipe flow, a similar position was achieved by D'Arcy's equation, yet scores of additional years of experimentation and effort were put in to replacing D'Arcy by a more precise analytical tool (Ch. 7).

With the analysis of open channel flows, no such effort has been attempted, simply because the velocity variation in open channels is so much greater (Fig. 11.4) than it is in manufactured pipes.

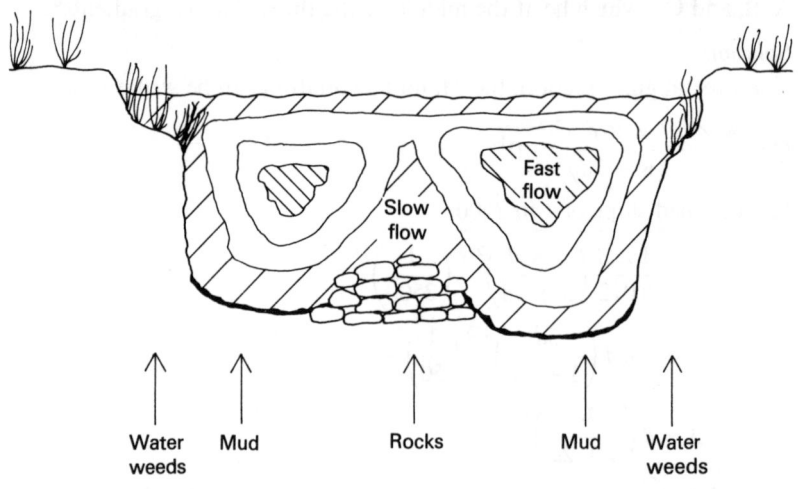

(a)

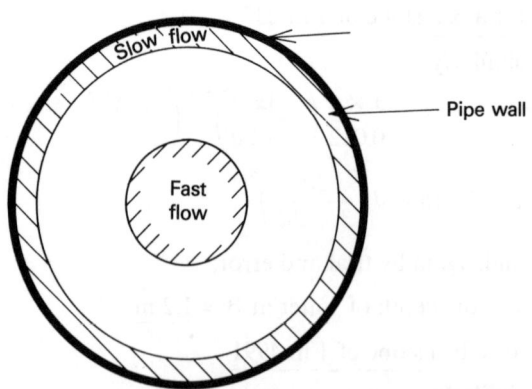

(b)

Fig. 11.4 Contours of velocity in a river's cross-section compared to those in a pipe. (a) Channel (b) Pipe

The velocity head in an open channel can be found apparently easily by use of a bent tube (Pitot tube). This operates by allowing fluid to enter its tip to lose its velocity on meeting the walls of the tube. Because energy can only be changed to some other type, and not lost, the velocity head is converted to pressure energy and the depth of the water in the tube stands higher than in the surrounding channel. The velocity head is calculated by a simple subtraction of these two depths (Fig. 11.5).

However, to carry out such an exercise across an open channel, at a number of levels, and then to calculate the precise average velocity of flow is a time-consuming and expensive procedure (and one that would have to be repeated for a large number of cross-sections because of the geometric variation found in natural channels) and is seldom worth

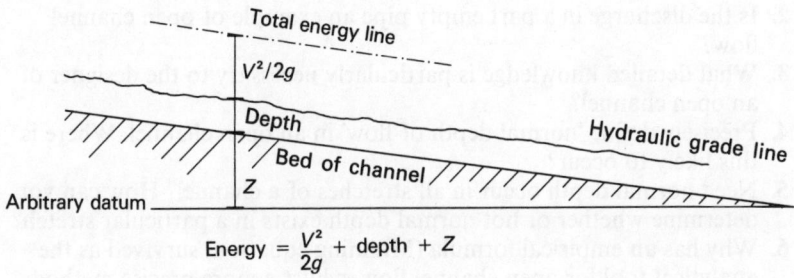

(a)

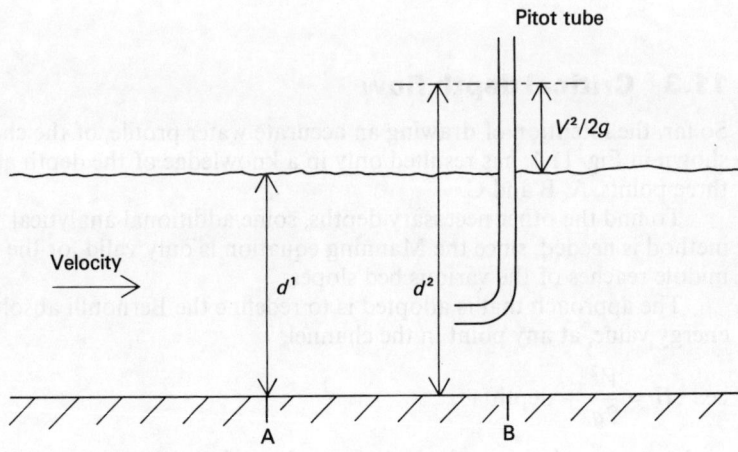

(b)

Fig. 11.5 Total energy and hydraulic grade lines in open channel flow. (a) Bernoulli terms for open channel flow (b) Measurement of velocity head ($v^2/2g = d_2 - d_1$)

considering. Instead, a less precise estimate of the average velocity is calculated from the measured discharge rate divided by the average cross-sectional area. Since the accuracy of measuring open channel flows is itself seldom more precise than ± 5 per cent (see Ch. 12), the average velocity of flow is thus a relatively inexact value.

Given that the average velocity, a term important in any fluid flow formula, is ill-defined, there is no practical reason to seek to improve on Manning's equation, which remains the basic equation of open channel flow.

Sections 11.1 to 11.2 – Self assessment questions

1. Concisely summarise the differences between open channel and pipe flow.
2. Is the discharge in a part empty pipe an example of open channel flow?
3. What detailed knowledge is particularly necessary to the designer of an open channel?
4. Precisely define 'normal depth of flow' in an open channel. Where is this likely to occur?
5. Need normal depth occur in all stretches of a channel? How can you determine whether or not normal depth exists in a particular stretch?
6. Why has an empirical formula (Manning equation) survived as the analytical tool for open channel flow, whilst a more precise method (Hydraulics Research Station charts) has superseded the empirical formula developed for pipe flow analysis?

11.3   Critical depth flow

So far, the intention of drawing an accurate water profile, of the channel shown in Fig. 11.1, has resulted only in a knowledge of the depth at the three points, A, B and C.

To find the other necessary depths, some additional analytical method is needed, since the Manning equation is only valid for the middle reaches of the various bed slopes.

The approach that is adopted is to redefine the Bernoulli absolute energy value, at any point in the channel,

i.e. $H = \dfrac{V^2}{2g} + \text{depth} + z$

as the energy relative to the bed of the channel at that point,

i.e.   specific energy at any point $= \dfrac{V^2}{2g} + d$ [11.4]

It is now interesting to see how the specific energy varies with the depth of flow.

In a laboratory channel – with a constant discharge rate passing through it – this is easily done by altering the channel slope to change the water depth. The results of such an experiment will be identical to those on Fig. 11.6, and it will be noticeable that every specific energy (with one single exception) can give rise to two types of flow –

      slow and deep flow,

or,   fast and shallow flow.

The sole exception is the point of **critical depth**, where the minimum value of specific energy needed to pass the flow occurs (point D – Fig. 11.6).

Since the specific energy (S.E.) at D is a minimum value, a differentiation of the specific energy at this point, with respect to depth of flow, has to equal zero,

i.e. $\text{S.E.} = \dfrac{V^2}{2g} + d$

$\qquad\quad = \dfrac{q^2}{2gd} + d$

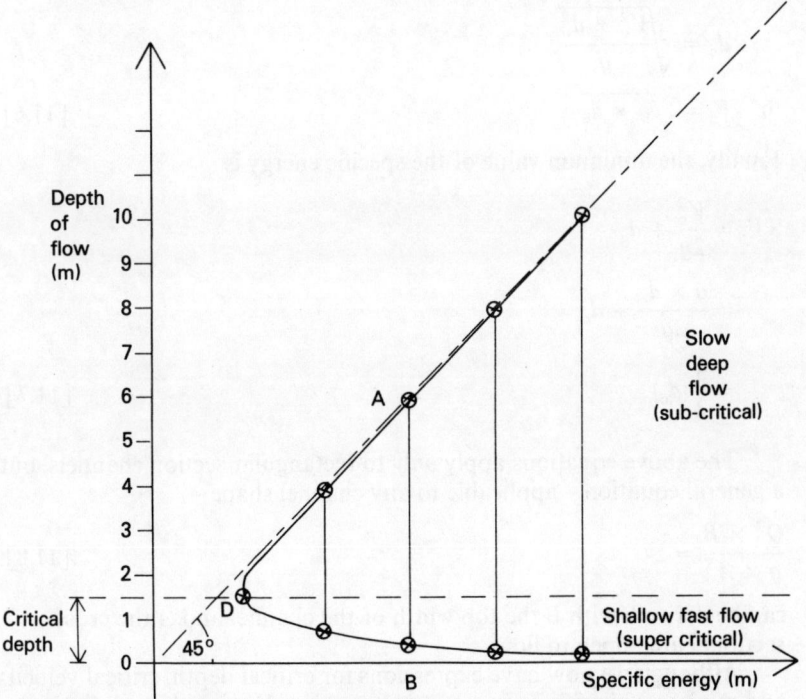

Fig. 11.6 Variation of specific energy with depth for a constant discharge rate

$$\left(\text{since} \qquad q = \text{the flow per metre width of channel} \right.$$

$$\left. \text{and so,} \qquad V = \frac{Q}{A} = \frac{q}{d} \right)$$

and

$$\frac{d \cdot (\text{S.E.})}{d \cdot d} = \frac{q^2}{2g}(-2d^{-3}) + 1$$

$$= 0$$

or

$$\frac{2q}{2gd^3} = 1$$

or

$$d = \text{the critical depth } (dc) = \sqrt[3]{\frac{q^2}{g}} \qquad [11.5]$$

The velocity which accompanies this depth is termed the **critical velocity** and can be obtained from

$$= V_c \times d_c$$

and substituting this value of q in [11.5] gives

$$d_c = \sqrt[3]{\frac{V_c^2 \times d_c^2}{g}}$$

or $V_c = \sqrt{g \times d_c}$ $\qquad [11.6]$

Finally, the minimum value of the specific energy is

$$\text{S.E.} = \frac{V_c^2}{2g} + d_c$$

$$= \frac{g \times d_c}{2g} + d_c$$

$$= \frac{3}{2}(d_c) \qquad [11.7]$$

The above equations apply *only* to rectangular section channels, but a general equation – applicable to any channel shape –

$$\frac{Q^2 \times B}{g \times A^3} = 1 \qquad [11.8]$$

can be derived, with B the top width of the channel and A the cross-sectional area open to flow.

Although we now have expressions for critical depth, critical velocity and minimum specific energy in a rectangular channel, the practical significance of these terms is still not obvious.

To clarify what they mean requires yet another experiment.

Across the floor of the laboratory channel a low mound – a **broad crested weir** – is built (Fig. 11.7) and the water is then allowed into the channel.

The initial water profile that results will be as shown in Fig. 11.7(a). This profile is quite understandable. The weir reduces the area open to flow and so the velocity of the water over it has to be faster. As any gain in velocity head is always matched by a decrease in the pressure energy,

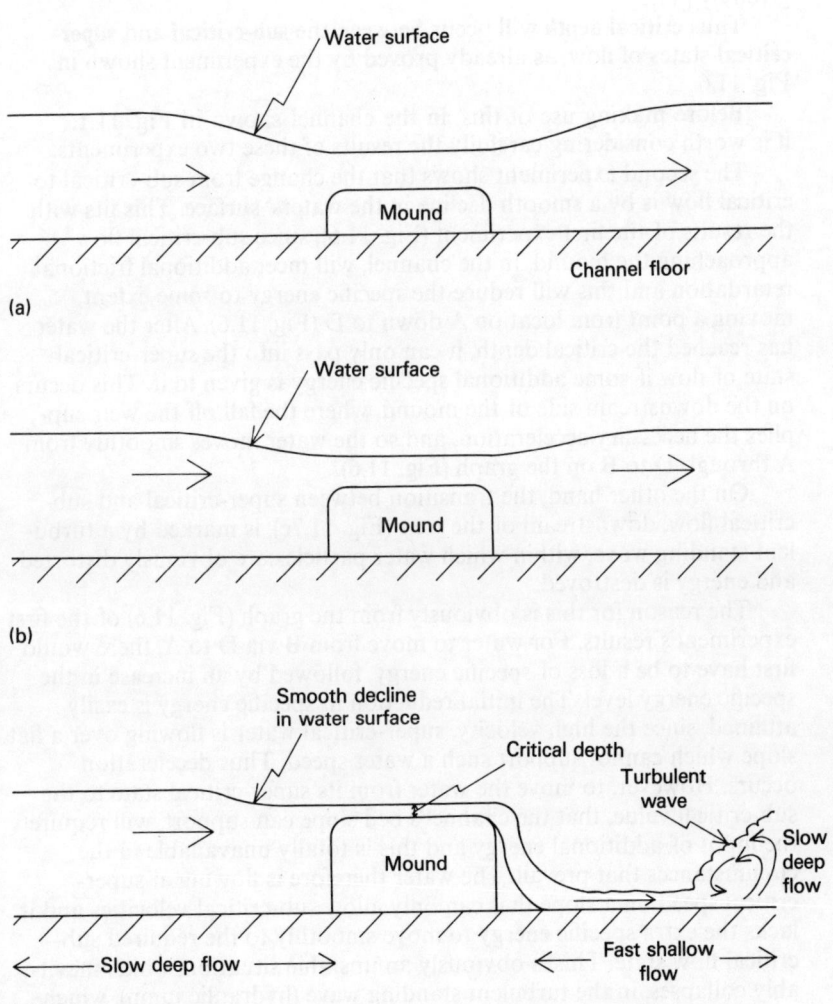

(a)

(b)

(c)

Fig. 11.7 The effect of increasing the height of a mound on a channel's floor. (a) Low mound (b) Somewhat higher mound (c) Still higher mound

the depth of water naturally falls above the mound. On the downstream side, the water, of course, returns to its original velocity and depth.

If the weir is now made a little higher, the profile in Fig. 11.7(b) is likely to occur. The greater drop in the water surface above the mound merely reflects the greater velocity of flow above the mound.

However, if the weir is made yet higher (Fig. 11.7c), a quite different profile will result. Upstream of the mound, slow deep flow will occur, whilst downstream will be fast and shallow water. A measurement of the water depth above the obstruction will reveal a depth identical to that given by [11.5].

Thus **critical depth** will occur between the **sub-critical** and **super-critical** states of flow, as already proved by the experiment shown in Fig. 11.6.

Before making use of this, in the channel shown in Fig. 11.1, it is worth considering carefully the results of these two experiments.

The second experiment shows that the change from sub-critical to critical flow is by a smooth decline in the waters' surface. This fits with the results of the first experiment (Fig. 11.6), since sub-critical flow approaching the mound, in the channel, will meet additional frictional retardation and this will reduce the specific energy to some extent, moving a point from location A down to D (Fig. 11.6). After the water has reached the critical depth, it can only pass into the super-critical state of flow if some additional specific energy is given to it. This occurs on the downstream side of the mound, where the fall off the weir supplies the necessary acceleration, and so the water moves smoothly from A through D to B on the graph (Fig. 11.6).

On the other hand, the transition between super-critical and sub-critical flow, downstream of the weir (Fig. 11.7c), is marked by a turbulent standing wave, within which water particles are obviously distorted and energy is destroyed.

The reason for this is obviously from the graph (Fig. 11.6) of the first experiment's results. For water to move from B via D to A, there would first have to be a loss of specific energy, followed by an increase in the specific energy level. The initial reduction in specific energy is easily attained, since the high velocity, super-critical water is flowing over a flat slope which cannot support such a water speed. Thus deceleration occurs. However, to move the water from its super-critical state to the sub-critical value, that the channel's bed slope can support, will require the input of additional energy and this is totally unavailable in the circumstances that prevail. The water therefore is flowing at super-critical speed on a slope that can only allow sub-critical velocities and it lacks the extra specific energy to move smoothly to the required sub-critical flow state. This is obviously an unstable situation, which inevitably collapses, in the turbulent standing wave (hydraulic jump), which invariably makes the transition from super-critical to sub-critical flow.

To return to the problem of Fig. 11.1's channel.

Example 11.2

Given that the flow rate in the 3-m wide, rectangular channel (Fig. 11.1) is 15 m³/s,

(a) what is the value of the critical depth of flow?
(b) what significance has this value, compared to the already calculated normal depths at points A, B and C?

Solution

(a) From [11.5],

the critical depth of flow $= \sqrt[3]{\dfrac{q^2}{g}}$,

where q the flow per metre width of channel

i.e. is $\dfrac{15}{3} = 5$ m³/s per metre width

$\therefore \qquad dc = \sqrt[3]{\dfrac{5 \times 5}{9.81}} = 1.336$ metres

(b) the calculated normal depths of flow are,

at A = 3 m
at B = 1.2 m
at C = 4 m

Thus the flow

on the **upper slope** of the channel is **sub-critical**,
on the **mid slope** of the channel is **super-critical**,
and, on the **lower slope** of the channel is **sub-critical**.

Critical depth values can thus be inserted at points D and E and the water profile from A to D to B will be a smooth **downdraw curve** (i.e. a curve marking out a decrease in the water's depth), whilst the water profile from B to E to C will be marked by a hydraulic jump at some point, as yet unlocated.

11.4 Downdraw and backwater curves

The water profile of Fig. 11.1's channel now has five points where the depth of water is adequately known. Between these points, however, there is still uncertainty.

Dealing with the transition from the sub-critical flow to the super-critical flow first (i.e. A to D to B), we know (from Fig. 11.7) that this will be a smooth downdraw curve, without any abrupt losses of energy.

Situations where gradual changes in velocity and pressure head take place are, of course, ideal for analysis by a Bernoulli approach (Ch. 6).

Consider Fig. 11.8, which shows a short length of the rectangular section channel with a bed slope of S (where $S = \tan \alpha$) and an energy line slope of i. Whether the water profile is that of a downdraw curve (where the depth decreases downstream) or of a backwater curve (where the depth increases) makes no difference to the following analysis, or to the resulting equation.

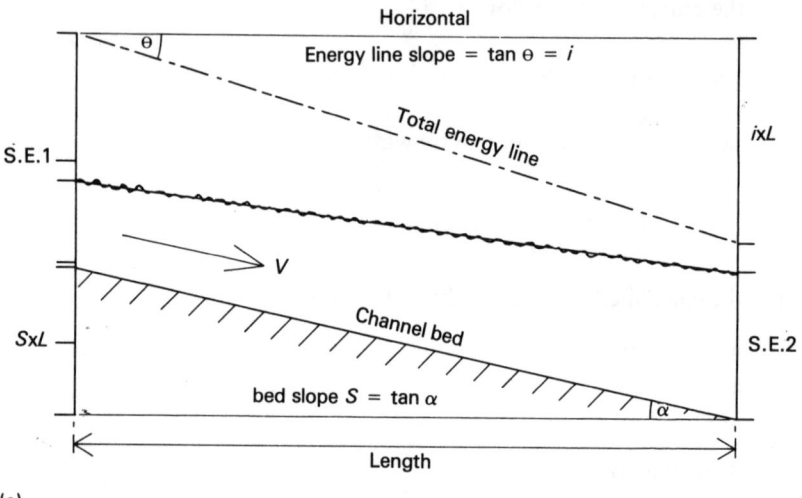

(a)

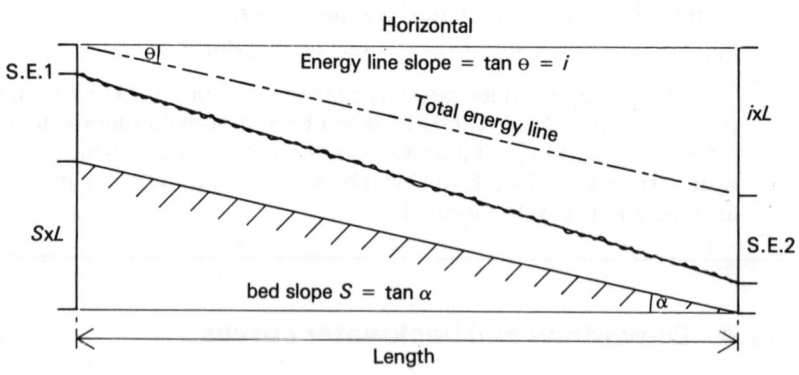

(b)

Fig. 11.8 Analysis of back-water and downdraw curves. (a) Backwater curve (b) Downdraw curve

The total energy, in Bernoulli terms, at the start of this length of channel is

(specific energy)$_1$ + (bedslope)(channel length) = S.E.$_1$ + $S \times L$

and at the outward end of the channel it is

(specific energy)$_2$ + (energy line slope)(channel length) = S.E.$_2$ + $i \times L$

Since these two energies can be seen to be equal, from Fig. 11.8, then

$$(S.E.)_1 + S \times L = (S.E.)_2 + i \times L$$

or

$$(S.E.)_1 - (S.E.)_2 = (i - s)L$$

But specific energy $= \dfrac{V^2}{2g} + d$

so $\left(\dfrac{V^2}{2g} + d\right)_1 - \left(\dfrac{V^2}{2g} + d\right)_2 = (i - s)L$

Whilst the term S is the fixed value for the channel's bed slope, the value i is the slope of the energy line and has to be calculated from the Manning equation, which can be rewritten as

$$i = \left(\frac{\bar{V} \times n}{\bar{m}^{2/3}}\right)^2 \tag{11.9}$$

The terms $\bar{V}$ and $\bar{m}$ are the average values of the velocity and the hydraulic radius, at the start and end of this length of channel.

Thus the downdraw curve equation can be rewritten as

$$L = \frac{\left(\dfrac{V_1^2 - V_2^2}{2g}\right) + (d_1 - d_2)}{i - s} \tag{11.10}$$

and, if the depths of and velocities at the start and end of the channel section are known, the length of the water profile can be calculated.

This equation is normally applied by taking two known sets of conditions (in the case of Fig. 11.1's channel, the normal depth and velocity at A and the critical depth and velocity at D), splitting the depth difference into as small increments as required and calculating the water profile.

An example will make the procedure more apparent.

Example 11.3
Determine the water profile between points A and D (Fig. 11.1), knowing that at A a normal depth of 3 m exists, and at D the critical depth is 1.366 m.

Table 11.2

(1) Depth (m)	(2) Velocity (m/s)	(3) $\frac{V^2}{2g}$ (m)	(4) Specific energy (m)	(5) $i = \left(\frac{nv}{m^{2/3}}\right)^2$	(6) $SE_1 - SE_2$	(7) Average i value	(8) Average $i - S$	(9) Length (L) (m)
3	1.667	0.142	3.142	0.0004				
					0.224	0.00045	0.00005	4480
2.75	1.818	0.168	2.918	0.0005				
					0.214	0.00057	0.00017	1258.9
2.5	2.000	0.204	2.704	0.00063				
					0.202	0.00073	0.00033	612.1
2.25	2.222	0.252	2.502	0.00082				
					0.183	0.00097	0.00057	321.1
2.0	2.500	0.319	2.319	0.00111				
					0.153	0.00134	0.00094	162.8
1.75	2.857	0.416	2.166	0.00156				
					0.055	0.00172	0.00132	41.7
1.634	3.060	0.477	2.111	0.00187				

Total length
(m) $= 6,876.6$

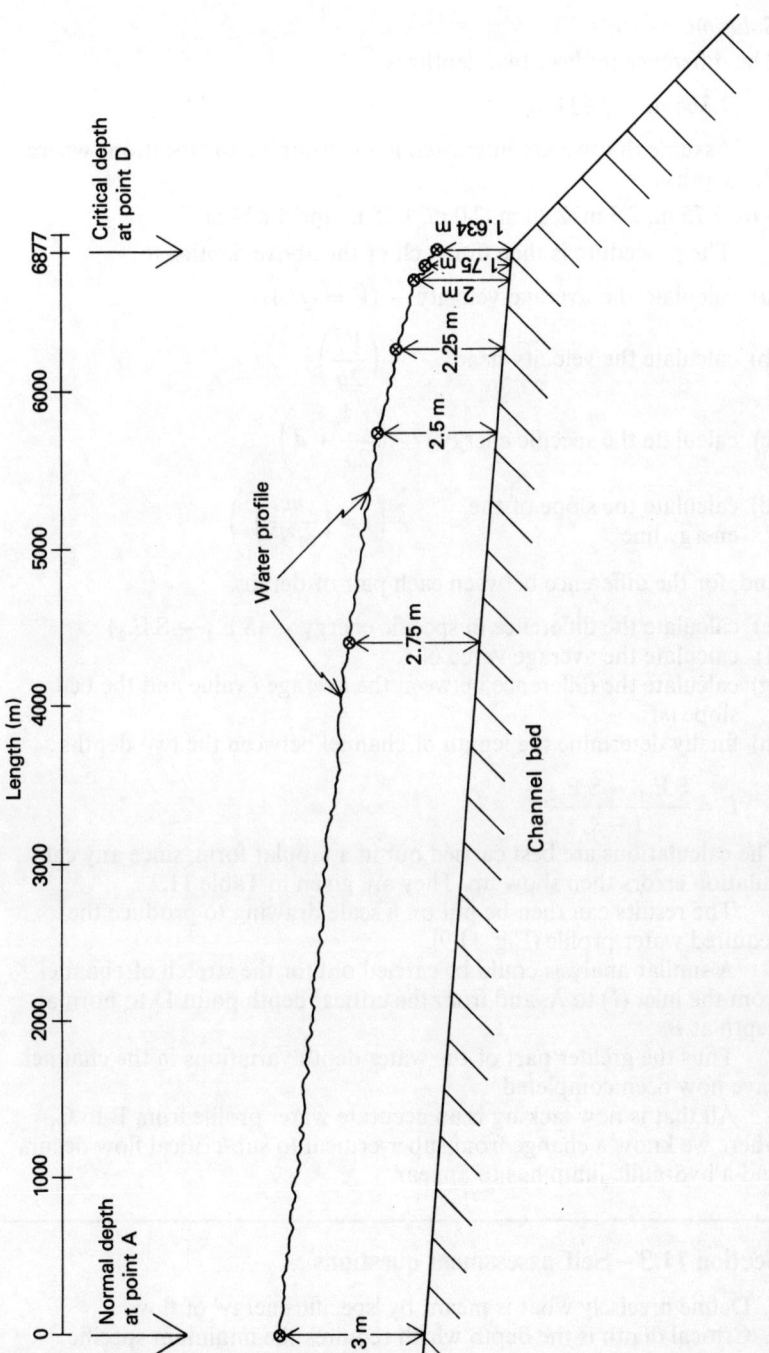

Fig. 11.9 Water profile between points A and D.

Solution

The difference in these two depths is

$$3 - 1.366 \text{ m} = 1.634 \text{ m}$$

Assume that we are interested in determining the locations where the depths are

3 m, 2.75 m, 2.5 m, 2.25 m, 2.0 m, 1.75 m and 1.634 m.

The procedure is then (for each of the above depths) to

(a) calculate the average velocity — $(\bar{V} = Q/A)$

(b) calculate the velocity head $-\left(\dfrac{V^2}{2g}\right)$

(c) calculate the specific energy $-\left(\dfrac{V^2}{2g} + d\right)$

(d) calculate the slope of the energy line $-\left(i = \left(\dfrac{nv}{m^{2/3}}\right)^2\right)$

and, for the difference between each pair of depths,

(e) calculate the difference in specific energy — $(\text{S.E.}_1 - \text{S.E.}_2)$
(f) calculate the average value of i
(g) calculate the difference between the average i value and the bed slope (s)
(h) finally determine the length of channel between the two depths

$$L = \frac{\text{S.E.}_1 - \text{S.E.}_2}{i - s}$$

The calculations are best carried out in a tabular form, since any calculation errors then show up. They are given in Table 11.2.

The results can then be put on a scale drawing to produce the required water profile (Fig. 11.9).

A similar analysis could be carried out for the stretch of channel from the inlet (I) to A, and from the critical depth point D to normal depth at B.

Thus the greater part of the water depth variations in the channel have now been completed.

All that is now lacking is an accurate water profile from B to C, where we know a change from super-critical to sub-critical flow occurs and a hydraulic jump has to appear.

Section 11.3 – Self assessment questions

1. Define precisely what is meant by 'specific energy' of flow.
2. Critical depth is the depth which requires the minimum specific energy, for the discharge rate required. Why then should channels never be designed to flow at critical depth?

3. Is the equation

$$dc = \sqrt[3]{\frac{q^2}{g}}$$

applicable to all channel geometries?

4. Define precisely the terms 'sub-critical flow' and 'super-critical flow' and explain what happens when either of these flow states changes to the other?

5. Why must a super-critical flow, entering a fairly flat channel slope, change to a sub-critical flow by means of a zone of disturbed and turbulent water?

6. Define precisely the terms 'downdraw curve' and 'backwater curve'.

7. Does the same analytical expression describe both downdraw and backwater curves?

8. In Example 11.3, the length of the downdraw curve between points A and D (Fig. 11.1) was required. Why was it necessary to calculate the channel lengths between six different pairs of depths?

11.5 The hydraulic jump

The presence of large turbulent eddies in the hydraulic jump has already been mentioned (section 11.3), and, as with pipe flow minor losses (Ch. 7), such eddies are marked by an abrupt loss of fluid energy. Thus, the energy conditions across a jump are as shown on Fig. 11.10.

Since there is an abrupt change in energy, the hydraulic jump cannot be analysed by Bernoulli methods. Instead, resort has to be made to equating the momentum and pressure forces across on both sides of the turbulent eddy.

The hydraulic jump is stationary, for any particular flow rate, at a point on the channel, thus the sum of the upstream forces must equal the sum of the downstream forces, i.e.

$$\rho \times g \times \frac{d_1^2}{2} + \rho q v_1 = \rho \times g \times \frac{d_2^2}{2} + \rho q V_2$$

where the terms of the type $\rho \times g \times \frac{d^2}{2}$

will be recognised as the average pressure intensity on a cross-section of water d metres high and one metre wide. Similarly, the terms of the type

$$\rho \times q \times V$$

will be recognised as the mass of water per metre width of the channel times the average velocity of the flow (i.e. the momentum of the water).

The above equation can be simplified to

$$d_1^2 - d_2^2 = 2q \frac{(V_2 - V_1)}{g}$$

202

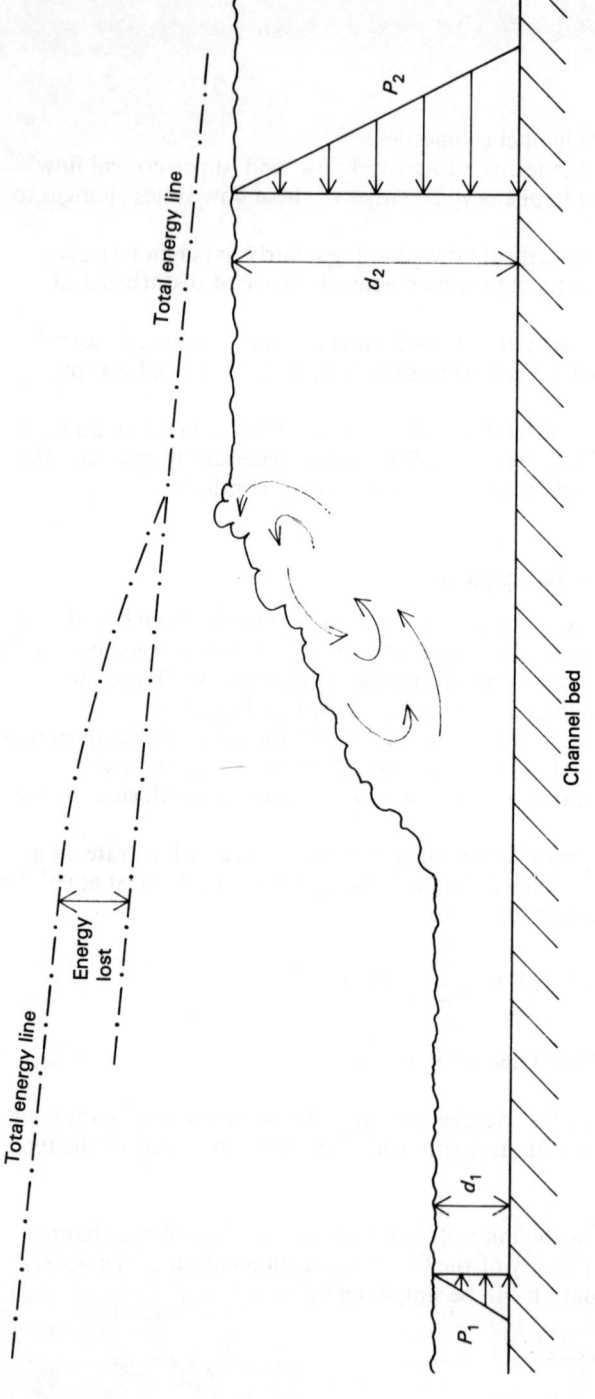

Fig. 11.10 Hydraulic jump in a rectangular channel

(1)

from which,

$$d_1 = -\frac{d_2}{2} \pm \sqrt{\frac{d_2^2}{4} + \frac{2q^2}{gd_2}}$$ [11.11]

or

$$d_2 = -\frac{d_1}{2} \pm \sqrt{\frac{d_1^2}{4} + \frac{2q^2}{gd_1}}$$

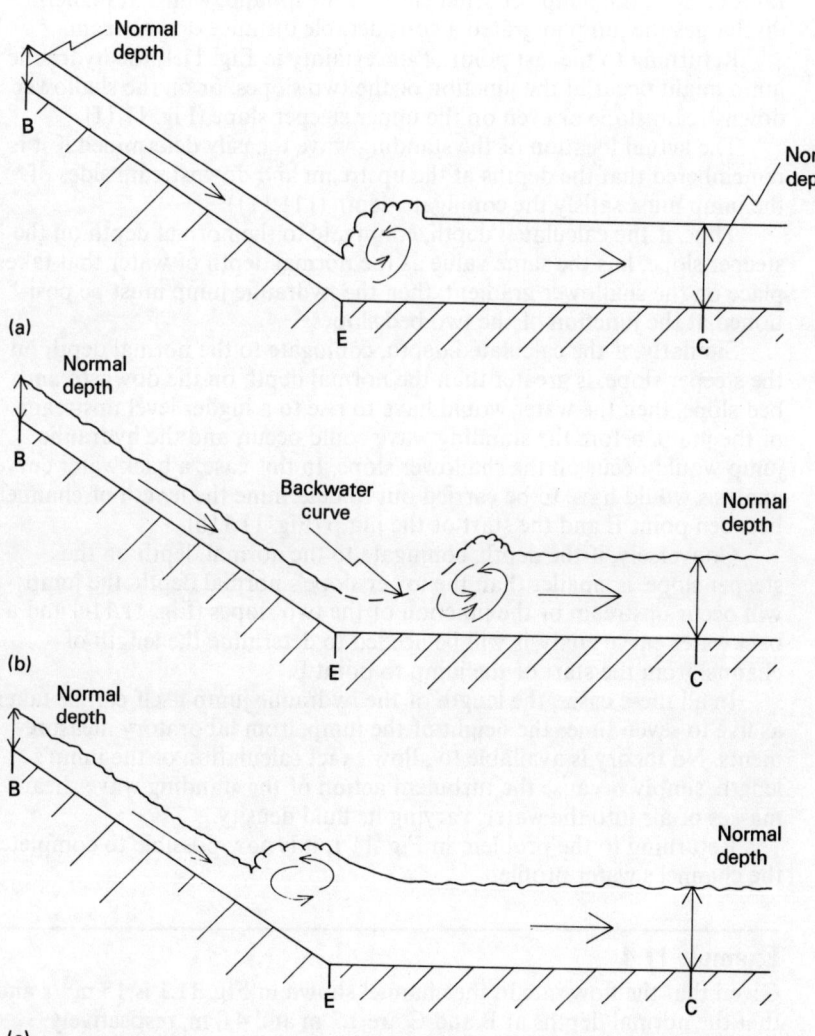

Fig. 11.11 Possible locations of the hydraulic jump in Fig. 11.1

As the negative values of the square root terms can be discarded as meaningless, a method is available to estimate the depth of water on one side of a hydraulic jump, provided that the depth of water on the other side is known. These depths are usually termed the **conjugate depths** in most standard texts.

If the experiment with the broad-crested weir in the laboratory channel (Fig. 11.7) had been carried out at a number of flow rates, it would have become apparent that the position of the hydraulic jump, downstream of the weir, changed as the flow rate was varied. With low rates of flow, the jump occurred close to the mound, whilst at higher discharges the jump migrated a considerable distance downstream.

Returning to the last point of uncertainty in Fig. 11.1, the hydraulic jump might occur at the junction of the two slopes, or on the shallower downstream slope or even on the upper steeper slope (Fig. 11.11).

The actual location of the standing wave is easily determined if it is remembered that the depths at the upstream and downstream sides of the jump must satisfy the conjugate depth ([11.11]).

Thus, if the calculated depth, conjugate to the normal depth on the steeper slope, has the same value as the normal depth of water that takes place on the shallower gradient, then the hydraulic jump must be positioned at the junction of the two bed slopes.

Similarly, if the calculated depth, conjugate to the normal depth on the steeper slope, is greater than the normal depth on the downstream bed slope, then the water would have to rise to a higher level upstream of the jump, before the standing wave could occur, and the hydraulic jump would occur on the shallower slope. In this case, a backwater curve analysis would have to be carried out to determine the length of channel between point E and the start of the jump (Fig. 11.11b).

Conversely, if the depth, conjugate to the normal depth on the steeper slope, is smaller than the lower slope's normal depth, the jump will occur upstream of the junction of the two slopes (Fig. 11.11c) and a backwater curve analysis will be needed to determine the length of channel from the start of the jump to point E.

In all these cases, the length of the hydraulic jump itself can be taken as five to seven times the height of the jump, from laboratory measurements. No theory is available to allow exact calculation of the jump's length, simply because the turbulent action of the standing wave shears masses of air into the water, varying its fluid density.

Returning to the problem in Fig. 11.1, it is now possible to complete the channel's water profile.

Example 11.4

Given that the flow rate in the channel shown in Fig. 11.1 is 15 m^3/s and that the normal depths at B and C are 1.2 m and 4.0 m, respectively, where will the hydraulic jump occur between the super-critical and sub-critical flow states?

Solution

The depth conjugate to the normal depth on the steeper slope (1.2 m) is (from [11.11])

$$d_2 = -\frac{d_1}{2} + \left(\frac{d_1^2}{4} + \frac{2q^2}{gd_1}\right)^{1/2}$$

$$= -\frac{1.2}{2} + \left(\frac{1.2 \times 1.2}{4} + \frac{2 \times 5 \times 5}{9.81 \times 1.2}\right)^{1/2}$$

$$= \underline{1.547 \text{ m}}$$

As the normal depth of flow on the lower bed gradient is greater than the calculated conjugate depth, the hydraulic jump must occur on the steeper channel slope (Fig. 11.11c).

To determine the exact location of the jump, all that is now necessary is to carry out a backwater curve analysis from the depth at the downstream side of the jump (calculated from [11.11]) to the downstream depth's normal flow (4 m).

When this is done, an accurate water profile for the channel of Fig. 11.1 can be drawn.

11.6 Summary of an open channel

Only a handful of analytical techniques are available for open channel flow –

- Manning's normal depth equation, the expressions describing critical depth, critical velocity, and specific energy, the backwater and downdraw iterative methods,
- the method of calculating conjugate depths for a hydraulic jump.

Provided that these techniques are applied where they are valid, the only point that need be borne in mind is that the likely behaviour of the water should be known and visualised **before any analysis at all takes place**. This is particularly important, since the behaviour of fluids in an open channel is outside of much of everyday experience, thus time spent in a laboratory seeing how water will behave in particular circumstances is time well spent.

Rectangular section channels have been used throughout this chapter to minimise the complexity of geometric calculations. However, the methods described can be applied to trapezoidal, semi-circular and to other regular channel shapes.

Additional worked examples are now given to widen the appreciation of open channel analysis.

Example 11.5

What flow can be expected in a 1.22 m wide rectangular channel (concrete lined) on a bed slope of 4 m in 10000 m, if the flow depth is 610 mm?

Solution

This is obviously a Manning type problem, and so

$$Q = \frac{A}{n} (m)^{2/3} (i)^{1/2}$$

$$= \frac{1.22 \times 0.610}{0.012} \left(\frac{1.22 \times 0.61}{1.22 + 2 \times 0.61}\right)^{2/3} \left(\frac{4}{10000}\right)^{1/2}$$

$$= 62.02 \times 0.453 \times 0.02$$

$$= \underline{0.562 \text{ m}^3/\text{s}}$$

Example 11.6

The flow in a rectangular culvert is measured as 0.412 m³/s, when the depth of flow is 0.500 m. The channel is 3.200 m wide and is laid on a bed slope of 1/2500. What is the roughness value of the channel lining material?

Solution

Again Manning's approach is required since,

$$Q = \frac{A}{n} (m)^{2/3} (i)^{1/2}$$

i.e. $n = \dfrac{A}{Q} (m)^{2/3} (i)^{1/2}$

$$= \frac{3.200 \times 0.500}{0.412} \left(\frac{3.2 \times 0.5}{3.2 \times 1.0}\right)^{2/3} \frac{1}{50}$$

$$= 3.883 \times 0.630 \times \frac{1}{50}$$

$$= \underline{0.049}$$

This is the usual way to determine the Manning number for channel materials, where an accuracy greater than that supplied by the published tables is wanted.

Example 11.7

A sewer pipe ($n = 0.015$) is laid on a slope of 0.0002 and is expected to carry the design flood flow (2.360 m³/s) when 90 per cent full. What diameter of pipe should be used?

Solution

A part full pipe is, of course, an open channel, since atmospheric pressure occurs above the surface of the fluid. Sewer pipes are usually designed to run part-full, at peak design discharge, to prevent any fluid rising up the manholes, and flooding the streets above.

As before, a Manning approach is appropriate and all that has to be noted is the different geometrical channel shape.

The hydraulic radius is calculated (Fig. 11.12):

$$m = \frac{A}{P} = \frac{\text{area of circle} - (\text{sector XOYA} - \text{triangle XOY})}{\text{circumference of circle} - \text{arc XAY}}$$

and since the angle $\alpha = \dfrac{1}{\cos \dfrac{0.4d}{0.5d}}$

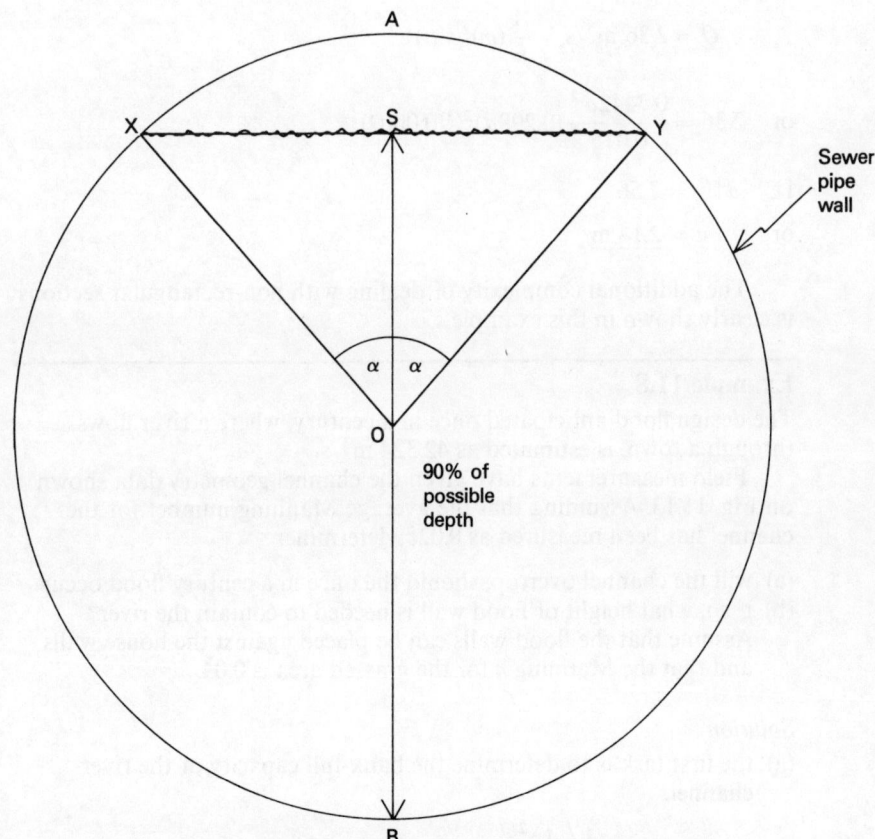

Fig. 11.12 Flow in a 90 per cent full sewer pipe (Example 11.7)

it equals 36° 52′

$$\therefore \quad \text{Area of sector XOYA} = 2 \times \left(\frac{36° \, 52′}{360°}\right)\left(\frac{\pi \times d^2}{4}\right)$$

$$= 0.1612d^2$$

$$\text{Area of triangle XOY} = 2 \times \frac{1}{2} \times 0.4d \times 0.4d \times \tan \alpha$$

$$= 0.120d^2$$

and the wetted perimeter $= \pi d - \left[\dfrac{2 \times 36° \, 52′}{360°}\right]\pi d$

(i.e. arc length XBY) $\qquad = 2.498d$

$$\therefore \quad m = 0.298d$$

$$\therefore \quad Q = 2.36 \text{ m}^3/\text{s} = \frac{A}{n}(m)^{2/3}\,(i)^{1/2}$$

or $\quad 2.36 = \dfrac{0.7442d^2}{0.015}(0.298d)^{2/3}(0.0002)^{1/2}$

i.e. $\quad d^{8/3} = 7.56$

or $\qquad d = \underline{2.14 \text{ m}}$

The additional complexity of dealing with non-rectangular sections is clearly shown in this example.

Example 11.8

The design flood anticipated once in a century, where a river flows through a town, is estimated as 42.324 m³/s.

Field measurements have given the channel geometry data shown on Fig. 11.13. Assuming that the average Manning number for the channel has been measured as 0.025, determine:

(a) will the channel overtop, should the once in a century flood occur?
(b) if so, what height of flood wall is needed to contain the river?

Assume that the flood walls can be placed against the house walls and that the Manning n for the grassed area is 0.04.

Solution

(a) the first task is to determine the bank-full capacity of the river channel.

i.e. $\qquad Q = \dfrac{A}{n}\left(\dfrac{A}{P}\right)^{2/3}(i)^{1/2}$

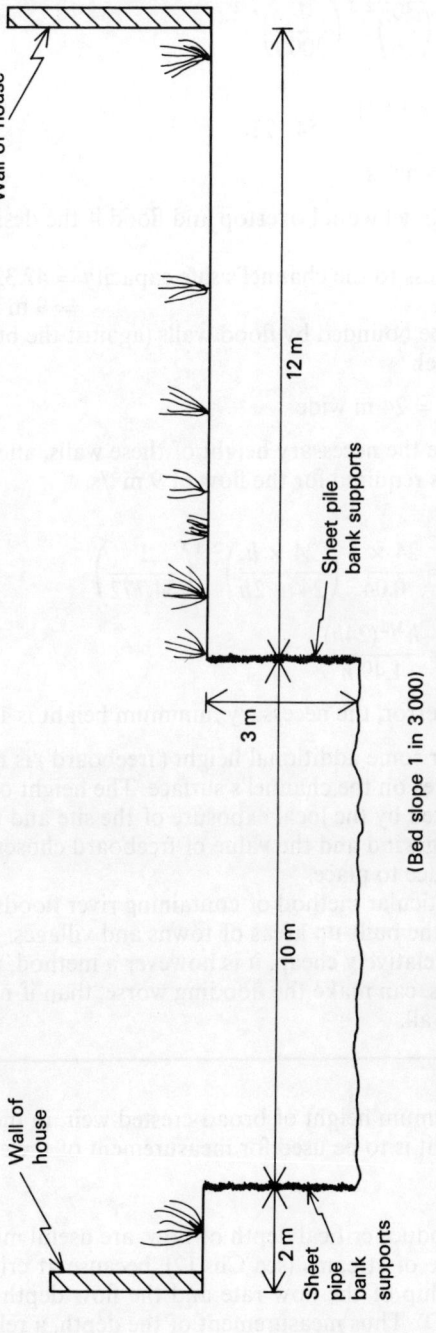

Fig. 11.13 Cross-section of channel through urban centre (Example 11.8)

$$\text{or } = \frac{30}{0.025}\left(\frac{30}{16}\right)^{2/3}\left(\frac{1}{3000}\right)^{1/2}$$

$$= 1200 \times 1.521 \times \frac{1}{54.772}$$

$$= 33.324 \text{ m}^3/\text{s}$$

Thus the channel would overtop and flood if the design flow occurred.

(b) The flow excess to the channel's safe capacity = 42.324 − 33.324
$$= 9 \text{ m}^3/\text{s}.$$

This has to be bounded by flood walls (against the buildings) in an upper channel:

$$2 + 10 + 12 = 24 \text{ m wide}$$

To determine the necessary height of these walls, another Manning calculation is required for the flow of 9 m³/s.
This is

$$9 = \frac{24 \times h}{0.04}\left(\frac{24 \times h}{24 + 2h}\right)^{2/3}\left(\frac{1}{54.772}\right)$$

$$\text{or } \quad 17.873 = \frac{h^{3/2}(24h)}{1.49\,h}$$

By trial and error, the necessary minimum height is 1.07 m.

However some additional height ('freeboard') is needed to contain any waves on the channel's surface. The height of waves is, of course, affected by the local exposure of the site and the direction of the prevailing wind and the value of freeboard chosen, thus, will vary from place to place.

This particular method of containing river floods is widely practised in the built-up areas of towns and villages. Although simple, and relatively cheap, it is however a method, which if any failure occurs, can make the flooding worse, than if no banks had been built at all.

___

Example 11.9

What is the minimum height of broad-crested weir, in the channel of Example 11.5, if it is to be used for measurement of the stream's flows?

Solution

Weirs, which produce critical depth of flow, are useful means of measuring the discharge of streams (see Ch. 12), because at critical depth a unique relationship of the flow rate and the flow depth exists (see Fig. 11.6, point D). Thus measurement of the depth, a relatively easy task, will provide the required flow rate.

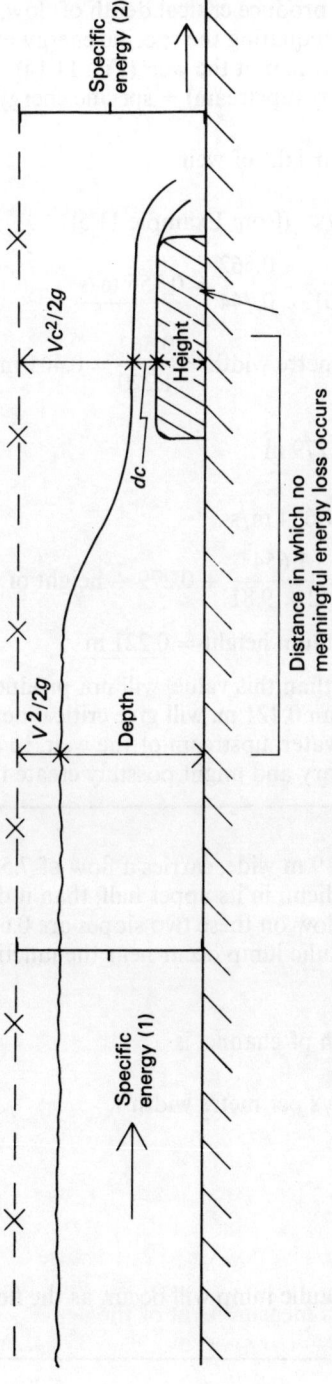

Fig. 11.14 The minimum height of a weir producing critical depth of flow

As the weir has to produce critical depth of flow, its minimum height is obtainable by equating the specific energy of a point, just upstream of the weir, to that at the weir (Fig. 11.14).

Since specific energy (upstream) = specific energy at weir

$$\frac{V^2}{2g} + d = \frac{Vc^2}{2g} + dc + \text{height of weir}$$

[since $Q = 0.562 \text{ m}^3/\text{s}$ (from Example 11.5)

$$V = \frac{0.562}{1.22 \times 0.61} = \frac{0.562}{0.744} = \underline{0.755 \text{ m/s}}$$

and $\quad q = \text{flow per metre width} = \frac{0.562}{1.220} = 0.461 \text{ m}^3/\text{s/m}$

and $\quad dc = \sqrt[3]{\frac{q^2}{g}} = 0.279 \text{ m}$

and $\quad Vc = \sqrt{gdc} = 1.654 \text{ m/s}$

$$\therefore \quad \frac{0.755^2}{2 \times 9.81} + 0.61 = \frac{1.654^2}{2 \times 9.81} + 0.279 + \text{height of weir}$$

and so, the weir's minimum height = $\underline{0.221 \text{ m}}$

A weir, any lower than this value, will not produce critical depth of flow. A weir, higher than 0.221 m, will give critical depth but will also increase the depth of water, upstream of the weir, to a depth greater than absolutely necessary and might possibly create flooding problems.

Example 11.10

A rectangular channel, 9 m wide, carries a flow of 7.500 m³/s and has a rather steeper bed gradient, in its upper half, than it does downstream. If the normal depths of flow on these two slopes are 0.614 and 0.853 respectively, will a hydraulic jump occur near the junction of the slopes?

Solution

The flow per unit width of channel is

$$= \frac{Q}{9} = \frac{7.5}{9} = 0.833 \text{ m}^3/\text{s per metre width}$$

and the critical depth,

$$dc = \sqrt[3]{\frac{q^2}{g}} = \underline{0.414 \text{ m}}$$

Therefore no hydraulic jump will occur, as the flow on both slopes is sub-critical.

Example 11.11

(a) What channel slope (using the data from Example 11.10) would be needed to produce critical flow depth? Assume that the Manning number is 0.012.

(b) Would it be wise to design a channel to flow at critical depth?

Solution

(a) From the Manning equation

$$7.5 = \frac{0.414 \times 9}{0.012} \left(\frac{0.414 \times 9}{9 + 2 \times 0.414} \right)^{2/3} (i)^{1/2}$$

or $7.5 = 310.5 \times 0.524 \times (i)^{1/2}$

$\therefore \quad i = 0.00212$

or the bed slope required = <u>1 in 471.7</u>

(b) It would seem sensible to design channels to run at critical depth, since in this condition a minimum specific energy is needed to pass the required flow rate.

    However, Fig. 11.6 makes it clear that for a very small change in specific energy, at the critical depth point, large variations in water depth can result. This is obviously undesirable, as is the fact that whether the depth increases or decreases is not known.

Example 11.12

A 3 m wide rectangular channel is laid on a slope of $1/1\,600$ to pass $4.444 \text{ m}^3/\text{s}$ of water. The channel's Manning number is 0.012.

    If a sluice gate in the channel is lifted 0.200 m above the channel bed, what is the water profile upstream and downstream of the gate?

Solution

Sluice gates are simple structures to control the depth and velocity of water in a channel. Essentially they are wooden or metal walls spanning the channel and able to be raised or lowered by a screw system.

    In this case the opening below the gate is 0.200 m and the problem is to decide whether this will create super-critical, critical or sub-critical flow depth.

    A Manning calculation reveals that the normal depth of flow in this channel is 1.000 m and the critical depth

$$\left(dc = \sqrt[3]{\frac{q^2}{g}} \right)$$

is 0.607 m.

    Thus flow beneath the gate is **super-critical**.

    The required water profile will be as that shown on Fig. 11.15.

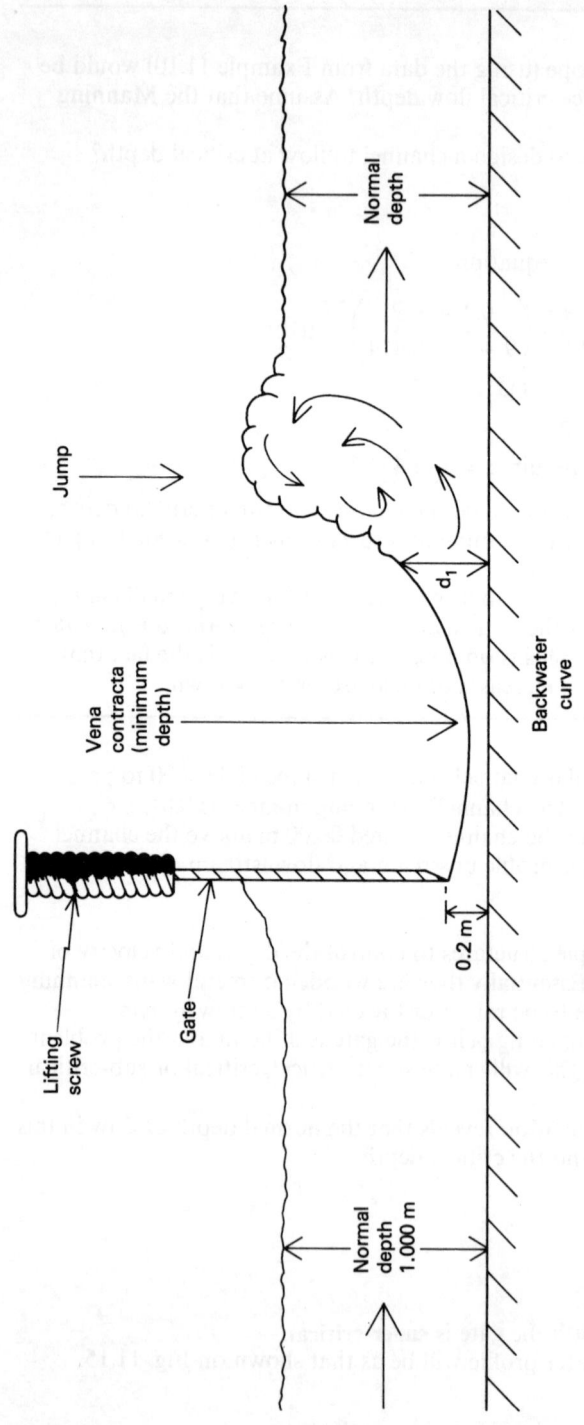

Fig. 11.15 Flow beneath a partly open sluice gate (Example 11.12)

On the upstream side of the gate the water depth will be slightly increased, because some velocity head is lost in contact with the gate.

Downstream, the escaping water reduces in depth from the 0.200 m value, below the gate, because of its inertia. The minimum water depth is thus at the end of this 'vena contracta'.

Downstream of this point, the water rises in a backwater curve until it reaches the depth conjugate to the normal depth for the channel bed slope and there a hydraulic jump occurs.

The downstream profile is the more complex and if it is necessary to calculate it in detail the procedure would be

(a) to use the conjugate depth equation to find the water depth before the jump (d_1);
(b) if a value for the vena contracta minimum depth can be estimated, the length of the backwater curve leading to the hydraulic jump can be calculated.

Example 11.13

How much energy is lost in the hydraulic jump (Example 11.12)?

Solution

Downstream of the jump, the water is at its normal depth (1 m) and so the average velocity of flow is

$$\frac{4.444}{3 \times 1} = \underline{1.481 \text{ m/s}}$$

and the velocity head $\left(\dfrac{V^2}{2g}\right) = \dfrac{1.481^2}{2 \times 9.81} = \underline{0.112 \text{ m}}$

Thus the specific energy downstream of the jump is

$$\frac{V^2}{2g} + d = \underline{1.112 \text{ m}}$$

Upstream of the jump, the water depth is the value conjugate to the downstream normal depth, i.e.

$$d_{\text{(upstream)}} = -\frac{d_2}{2} + \sqrt{\frac{d_2^2}{4} + \frac{2q^2}{gd^2}}$$

$$= \underline{0.335 \text{ m}}$$

Thus the velocity is $\dfrac{4.444}{3 \times 0.335} = \underline{4.421 \text{ m/s}}$

and the velocity head is $\underline{0.996 \text{ m}}$

The specific energy upstream of the jump is thus $\underline{1.331 \text{ m}}$

and the energy lost in the jump is

1.331 − 1.112 m = 0.219 m

Hydraulic jumps are deliberately caused where reservoirs discharge into streams to reduce the specific energy and thus the velocity of flow. If this were not done then the fast flow could cause a hazard to people and animals and could also cause erosion of the stream's banks (see section 11.8.3.)

11.7 The design of open channels

11.7.1 The cost of open channels

Open channels are invariably highly expensive works. Thus their design must take note of both the capital cost of construction and the recurrent maintenance charges which will have to be borne for the life of the channel.

11.7.2 The Best section concept

To reduce the capital cost, it is obviously sensible to build as small a channel as is technically possible, and this means reducing the necessary excavation to a minimum.

From Manning's equation, it is apparent that the maximum discharge rate – for a chosen bed gradient and channel lining roughness – occurs when the hydraulic radius (m) has its largest value. Or, since

$$m = \frac{A}{P}$$

this is achieved when the wetted perimeter (P) is as short as possible.

A semi-circle best meets this condition of maximum area with minimum perimeter length, but channels of semi-circular cross-section are difficult (and thus expensive) to build. So recourse is usually made to the simpler and cheaper rectangular or trapezoidal cross-sections.

For a rectangle, the minimum perimeter length can be established from

Area $A = W \times d$

Perimeter $P = W + 2d$

thus, for any value of A,

$$P = \frac{A}{d} + 2d$$

and the minimum value of this is found when its differential is made to equal zero.

i.e. $\dfrac{dP}{dd} = -\dfrac{A}{d^2} + 2 = 0$

or $\quad W = 2d$ [11.12]

Similarly, for a trapezoidal channel, the minimum perimeter length occurs when the channel's top width equals twice the length of a sloping side (S) (Fig. 11.16).

In both cases, the **best hydraulic section** is tangential to the truly most efficient shape, that of a semi-circle.

Although both these 'best' sections permit the maximum discharge rate, for the bed slopes and roughness adopted, it must be recognised that this is accompanied by the highest possible average velocity of flow for these conditions. This will make a 'best' section quite unsuitable for the majority of unlined open channels.

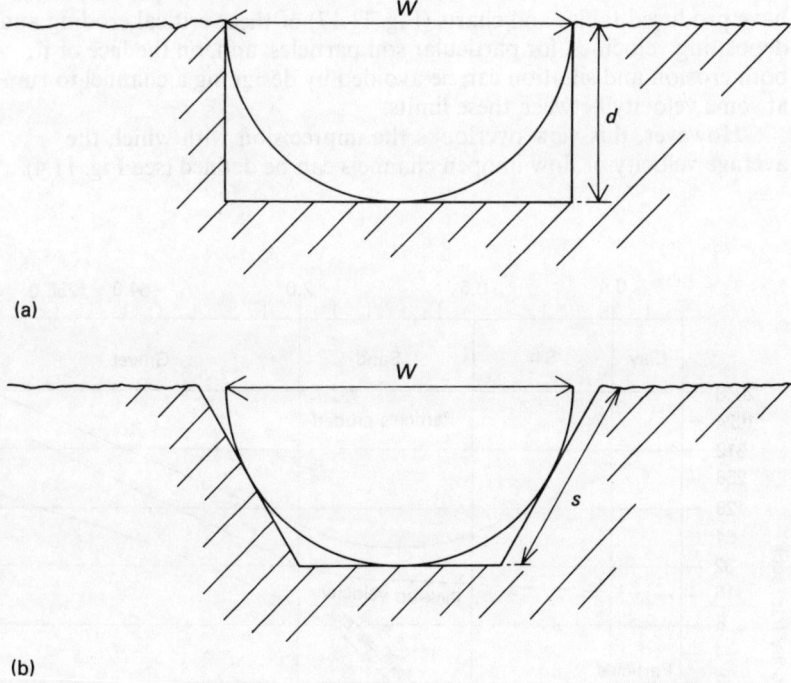

(a)

(b)

Fig. 11.16 Most 'efficient' cross-sections for open channels. (a) Rectangular cross-section ($W = 2 \times d$) (b) Trapezoidal ($W = 2 \times s$)

11.7.3 Lined and unlined channels

The choice of whether or not to line an open channel is sometimes removed from the designer's hands. The case of a water supply canal cut through porous and permeable sands is obviously an example of this. If the channel is left unlined, leakage through its boundaries would cause such a loss of flow that the canal would probably fail in its designed role.

In other cases, the choice is the designer's and, at this stage, the question of relative cost is considered. The capital cost of a lining, of whatever material, is easily established, but capital costs do not describe the real charge of a structure over years of use. The expense of maintenance has, of course, to be considered.

In open channels, maintenance largely depends on the ability of the water at one point to erode, pick up and transport particles from the channel walls, and at some downstream point – where the velocity of flow is less – to deposit the suspended particles and silt up the channel's section.

Research has shown that the transporting power of water varies with the sixth power of the fluid velocity, thus quite small variations in velocity can lead to significant increases in erosion. Years of experimentation have produced tables and charts (Fig. 11.17) of these critical eroding and depositing velocities, for particular soil particles, and, on the face of it, both erosion and siltation can be avoided by designing a channel to run at some velocity between these limits.

However, this view overlooks the imprecision with which the average velocity of flow in open channels can be defined (see Fig. 11.4).

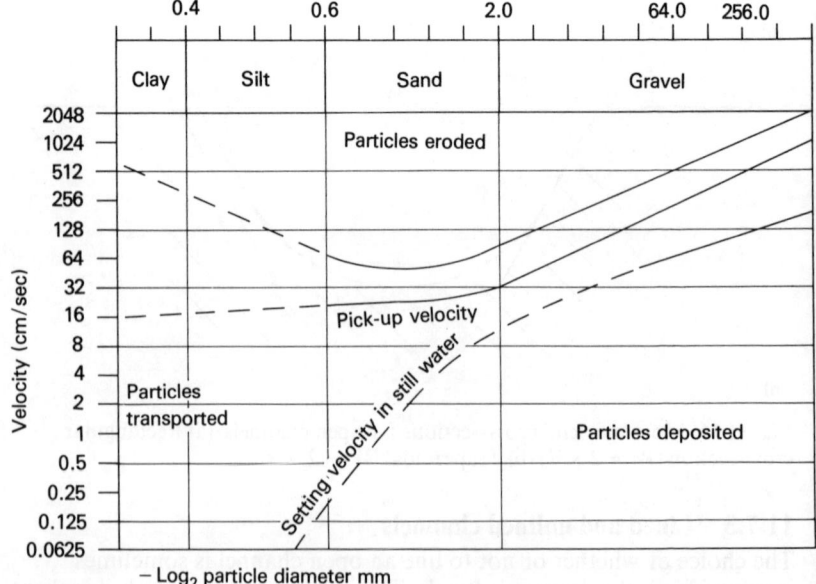

Fig. 11.17

Thus the tendency is to design two quite different types of channel:

(a) **non-eroding channels**, where no lining material is provided;
(b) **self-cleansing lined channels**.

With unlined channels, the worry is that high velocity flow will erode parts of the bank and allow the remainder to slump into the channel. This is usually avoided by designing such channels to run at velocities well below the erosion value. Thus the flow in unlined open channels tends to be slow and the cross-sectional area is invariably greater than that suggested by the 'best' section formulae ([11.12]).

On the other hand, whilst the worry of erosion can be ignored with lined channels, siltation can be a real problem. This is usually avoided by ensuring that the average velocity of flow is always above the 'pick-up' value (Fig. 11.17) for whatever solid materials are likely to wash into the channel. As the quantity of flow in an open channel will vary with local rainfall over its catchment, this of course means that for every possible depth of flow the velocity must be self-cleansing. Thus the flow in lined open channels runs at high velocities (so much so that access to them is often denied to the general public) and a cross-sectional width that reduces with flow depth is often essential. Many such channels tend to a triangular cross-section and so, yet again, the 'best' sectional dimensions are often unsatisfactory.

The usual design procedure is thus to determine the best hydraulic section and then to modify it, as necessary, to meet the requirements of either an unlined or lined channel. Often two designs – one lined and the other unlined – are prepared and costed, to ensure that the cheapest choice is made.

Example 11.14

An 8 km long channel is to be designed to pass flows of up to 15.000 m^3/s on a bed gradient of 1 in 2 500.

The soils on the channel's route are silty (Manning number 0.028) and will erode at water velocities in excess of 0.750 m/s. As silt particles will inevitably be washed into the channel in wet weather, the lowest velocity of flow must always exceed that (0.0013 m/s) needed to carry the particles.

A 50-year working life is expected of the channel and the financial factors of greatest interest are –

Excavation – £1.35 per m^3

Concrete lining – £4.10 per m^2

On the assumption that major erosion or siltation can be avoided by the design, the annual minor maintenance costs have been estimated as £63.75 per kilometre (lined channel) and £1400 per kilometre (unlined channel).

Choose a suitable channel section.

Solution

Both lined and unlined sections should be chosen and costed to produce as cheap a channel as possible.

(a) *Lined channel*. Here both rectangular and trapezoidal sections are possible.

Rectangular section
For the best hydraulic section (see section 11.8.2)

$$W = 2 \times d$$

Thus, with a concrete lining, whose roughness is described by a Manning number of 0.014, and the peak discharge rate, the Manning equation can be written as

$$15.000 = \frac{2d^2}{0.014}\left(\frac{2d^2}{4d}\right)^{2/3}\left(\frac{1}{2500}\right)^{1/2}$$

From which, a water depth of 2.2145 m and a channel width of 4.429 can be calculated.

These dimensions, whilst mathematically accurate, would present practical difficulties when the channel is being set out and constructed. It would be preferable, therefore, to adopt the easier constructed dimensions of

$$d = 2.500 \text{ m}$$
and $$W = 4.500 \text{ m}$$

If this revised width value is inserted in the Manning equation, a flow depth of 2.17 m for the 15.000 m^3/s flow rate is produced. Thus a useful freeboard of 0.33 m is available to contain surface waves.

As the channel is to be lined, the important maintenance problem is that of siltation and a check of the average velocity of flow at various possible depths (and thus discharge rates) shows the following:

Depth of flow m	Average flow velocity (m/s)
2.17	1.536
1.50	1.330
1.00	1.170
0.50	0.785

Therefore this section should always produce flow velocities, no matter the discharge rate, which will transport any silt particles out of the channel.

Thus the channel will operate satisfactorily and will carry the highest design discharge with an adequate freeboard.

Its costs can now be calculated as:

Excavation = £1.35 × 2.5 × 4.5 × 8 000
= £121,500

Concrete lining = £4.10 × (5.0 + 4.5) × 8 000
= £311,600

50 years of minor maintenance = £25,500

Total = £458,600

Trapezoidal section. A wide range of trapezoidal sections is possible, depending on the bottom width and the side slopes selected. Normally this choice is reduced to those which suit the available excavation plant, and the shutters for placing the concrete lining. For this example it will be assumed that a channel bottom width of 1 m and side slopes of 1 (vertical) in 2 (horizontal) are preferable, in the light of existing plant and equipment.

Before attempting to use the Manning equation, it is convenient to examine the geometry of a trapezium. If this is done, assuming 1 in 2 side slopes, the following geometric relationships can be found:

cross-sectional area $\quad(A) = b \cdot d + 2d^2$

wetted perimeter $\qquad(P) = b + 2d(5)^{1/2}$

length of sloping side $\quad(S) = d(5)^{1/2}$

top width $\qquad\qquad(W) = b + 4d$

where b is the channel's bottom width,

and d is the depth of flow

The Manning equation, for the maximum discharge rate, then produces a depth of flow of 2.036 m.

A costing of this trapezoidal section reveals that (at £468 475.60) it is rather more expensive than the chosen rectangular section. Thus time spent, either in modifying the dimensions, to ease setting out, or in checking that the channel will always operate in a self-cleansing manner, is unjustified.

(b) Unlined channel. Only a trapezoidal section is feasible, since unsupported vertical walls of the local soil would not be stable.

In practice, it is important to determine the slopes that are stable, both when the soils are wet and dry, and that usually entails field surveys and laboratory tests. For the purposes of this example, it will be assumed that slopes of 1 (vertical) in 2 (horizontal) will be suitable.

Using the bottom width, side slopes and the geometric relationships developed above, the Manning equation for the highest discharge produces a flow depth of 2.71 m, but with an average velocity of flow (0.862 m/s) well in excess of that which will scour the channel walls.

The section, therefore, has to be modified to increase the wetted perimeter and thus increase the frictional drag and reduce the average velocity to a more acceptable figure.

One way to achieve this is to keep the bottom width of 1 m and to flatten the side slopes to 1 (vertical) in 5 (horizontal). These revised side slope values of course modify the expressions for the cross-sectional area and the wetted perimeter to

$$A = d + 5d^2$$

and $P = 1 + 10.198d$

Use of the Manning equation then gives a flow depth of 1.94 m at the acceptable velocity of 0.714 m/s.

This section is unlined and so has to be non-eroding at all possible depths of flow. Insertion of various flow depths in the Manning equation gives:

Water depth m	Average velocity (m/s)
1.94	0.714
1.50	0.606
1.00	0.471
0.50	0.310

It therefore can be seen that the chosen section is non-eroding at all possible discharge rates.

To contain any surface waves, some freeboard should be provided. A figure of 0.36 m is probably sufficient, giving a channel 2.30 m deep and with bottom and top widths (respectively) of 1 and 24.0 m.

The costs of this channel section are:

$$\text{Excavation} = £1.35 \times 28.75 \times 8000$$
$$= £310\,500$$

$$\text{Minor maintenance over 50 years} = £1400 \times 8 \times 50$$
$$= £560\,000$$

$$\text{Total} = £870\,500$$

As this is approximately double the cost of that of the lined rectangular section, this unlined trapezoidal channel is obviously far too expensive, as well as occupying too great an area of land, with a channel top width of 24 m.

Conclusion

A best hydraulic section could have been adopted for the rectangular, concrete-lined channel, and this was only modified to make the construction task simpler. With the unlined channel, a top width rather greater than that for a best hydraulic section proved necessary, to ensure that a non-eroding channel resulted.

Many other possible sections could have been attempted, making the task of producing the most economic choice of section long and tedious. However, the costs of open channels are such that the design effort and costs needed, for the cheapest channel choice, is generally insignificant.

11.7.4 Non-uniformities in open channels

The above discussion implicitly assumed that Manning-type normal depth flow would occur throughout the channel.

However, non-uniform flow conditions do occur, even in quite short channels, and good design has to take note of this.

Apart from the earlier discussed effects of changes in the channel's bed slope, other common non-uniformities arise from bends, channel transitions and obstructions, such as bridge piers.

If a **bend** has to be included in the plan of a channel, the water flowing round it in a curved path is acted upon by a force (centripetal force) directed towards the centre of curvature of the bend. Opposing this is a centrifugal force. The result of this additional complexity of forces is that the depth, on the outer side of the channel, is always higher than on the inner side (Fig. 11.18).

Theoretically, this difference in depths is described by the expression

$$d_2 - d_1 = \frac{V^2 \times \text{channel width}}{g \times \text{radius of bend}} \qquad [11.13]$$

However, the variation in open channel velocity is particularly great around a bend and the above equation tends to under-estimate the difference in water depths. Field measurements of the velocity distribution (by use of a current meter – Ch. 12) around a bend of similar radius are normally required to fully evaluate the equation.

In all cases, however, additional height is required for the outer bank of a channel bend.

Channel transitions are no more than special sections which join lengths where the channel has, for some particular reason, different geometries. A common instance is where a channel has to be culverted under a roadway.

If the flow is sub-critical the problem is no more than avoiding abrupt changes in geometrical shape. Tapered boundaries are utilised to avoid setting up turbulent eddies with their associated head losses and scouring effects. With gradually tapered walls, the head losses can be reduced to between one-tenth and one-fifth of the incoming velocity head.

The variation in flow profile through a culvert can be large (Fig. 11.19), and particular design care is always needed at such conditions.

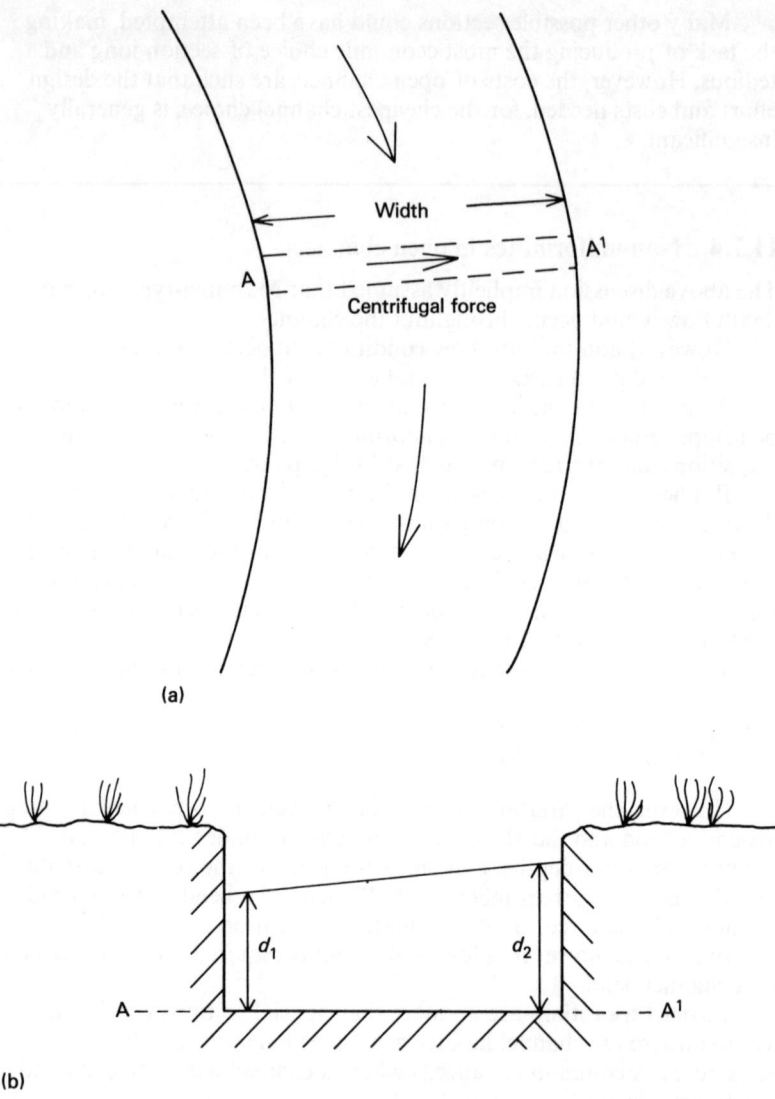

(a)

(b)

Fig. 11.18 Flow around a bend in an open channel. (a) Plan view (b) Section

Should it be necessary to pass super-critical flow through a channel transition, the problem is more serious, and use often has to be made of scale models of the channel transition, to ensure that uncontrolled hydraulic jumps do not raise the water level to the point where it will overtop the channel bank.

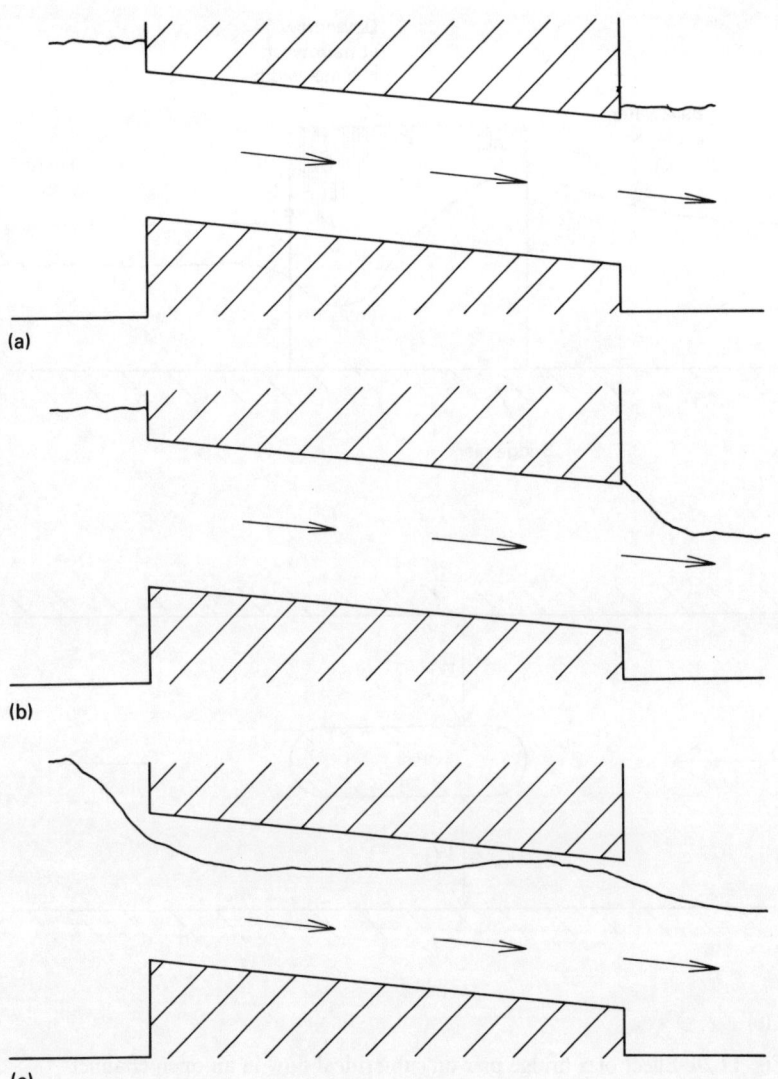

Fig. 11.19 Some possible profiles of flow through a culvert joining two sections of open channel

Obstructions such as **bridge piers** in an open channel give much the same effect as constrictions of the channel's area. In both cases a higher velocity of flow, and thus a lower depth, result. A common water profile, where sub-critical flow encounters a bridge pier, is shown in Fig. 11.20.

The subject of open channel obstructions has received a considerable research interest over the years and design charts, covering the

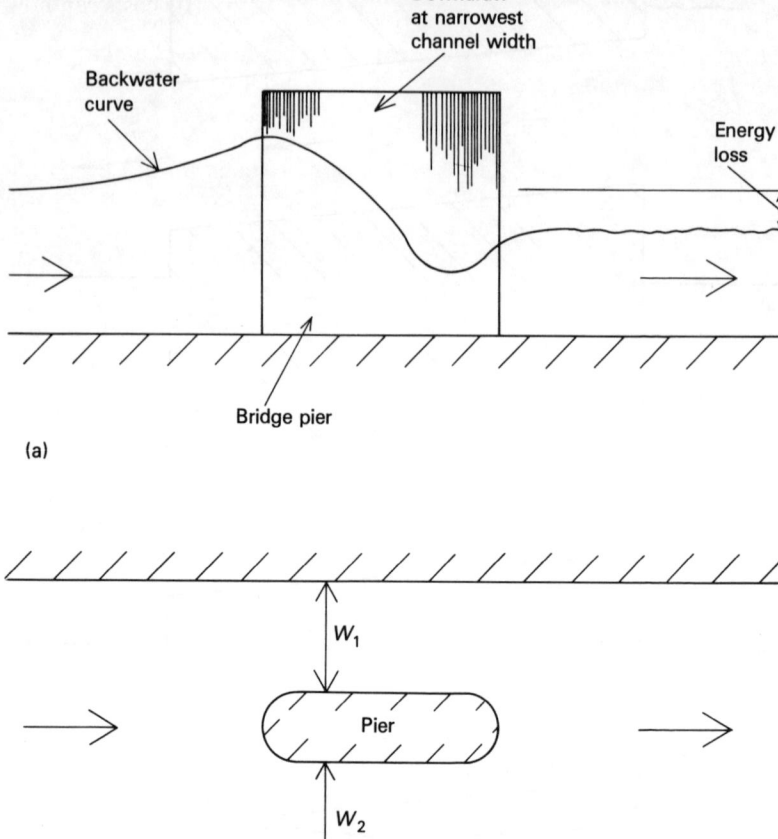

Fig. 11.20 Effect of a bridge pier on sub-critical flow in an open channel. (a) Profile of flow (b) Plan of obstruction

commoner cases, are given by Ven Te Chow, (*Open Channel Hydraulics*, McGraw-Hill Book Company 1959).

As the backwater effects caused by such obstructions can be transmitted a considerable length upstream, this is another area where particular design care is needed.

11.7.5 Summary

The choice of an economical channel width is quite simple, though it does involve a large amount of work checking that 'best' cross-sections meet the demands of either non-eroding or self-cleansing design. If channels could always be arranged to flow at normal depth conditions, then their design would not be a particularly complicated matter.

However, non-uniformities which cause a departure from the Manning flow state are common and present the designer with his greatest problems.

Only the simplest of such cases have been mentioned, yet even these are only precisely solvable if a regular channel geometry and sub-critical flow occur. In the more complex cases, recourse has to be made to scale-model analysis, and this normally implies the help of a specialist hydraulics laboratory.

Chapter 12

Measurement of fluid flow rates

12.1 Introduction

Measurements of the rate of fluid flow are often required routinely in both pipelines and in open channels.

In line with the differences already noted between pressure pipe and open channel flow, the methods employed are dissimilar. In general, the measurement of pipe flow is more precise, as well as easier and cheaper to carry out.

12.2 Measurement of pressure pipe flows

Pressure pipe flows are best determined by a Bernoulli analysis of the energy conditions at, and upstream of, a tapered length of pipe (Fig. 12.1).

Since the distance between the two piezometers is short, frictional head losses should be small, and the tapered change in pipe section should minimise the production of energy-degrading turbulent eddies. Thus, as a first approximation, the energy level at point (1) can be assumed to equal that at point (2), i.e.

$$\frac{V_1^2}{2g} + \frac{P_1}{\rho g} + Z_1 = \frac{V_2^2}{2g} + \frac{P_2}{\rho g} + Z_2$$

or, noting that $h = \dfrac{P}{\rho g}$,

$$\frac{V_2^2 - V_1^2}{2g} = (h_1 - h_2) + (Z_1 - Z_2)$$

229

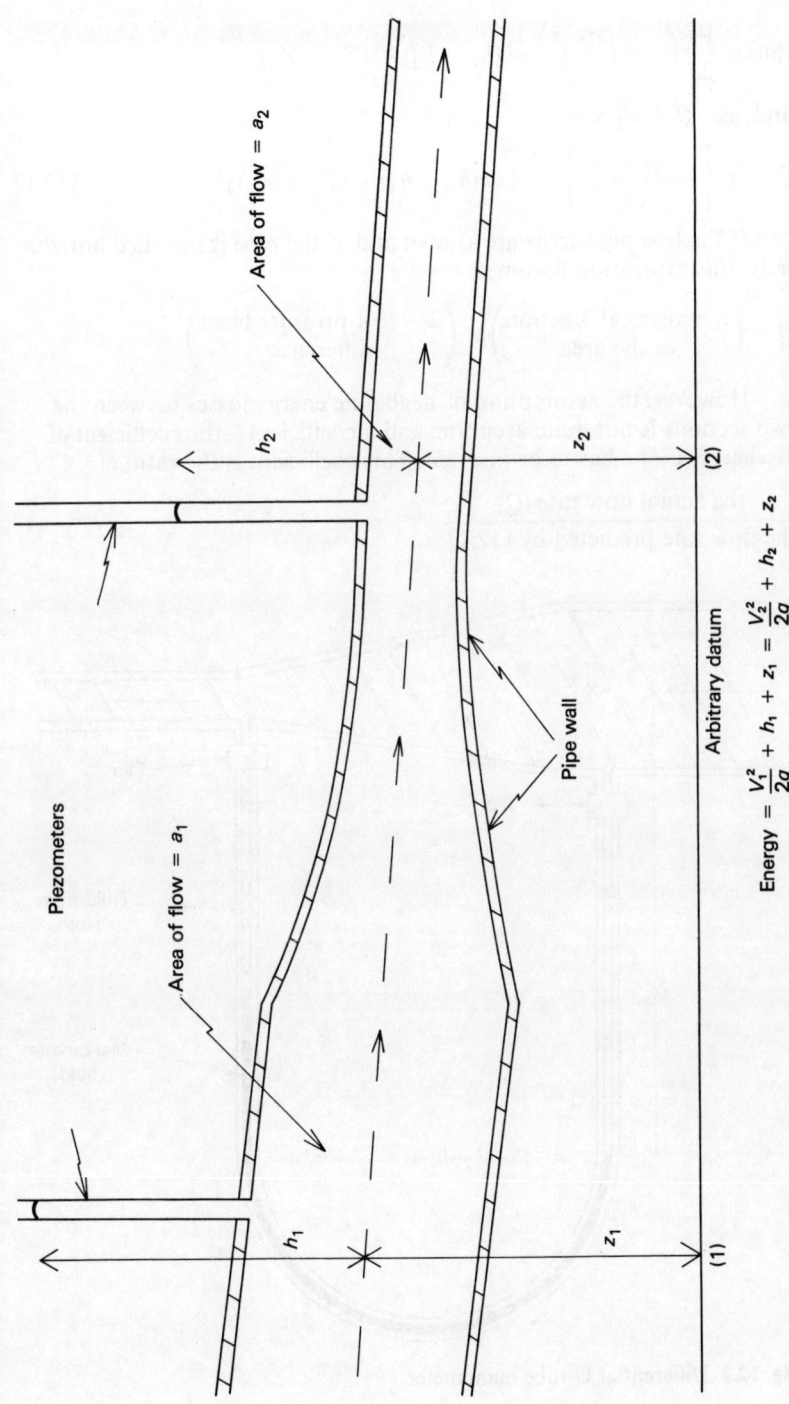

$$\text{Energy} = \frac{V_1^2}{2g} + h_1 + z_1 = \frac{V_2^2}{2g} + h_2 + z_2$$

Fig. 12.1 Basis of flow measurement in pressure pipes

thus, $\dfrac{V_2^2}{2g}\left(1-\left(\dfrac{a_2}{a_1}\right)^2\right)=(h_1-h_2)+(Z_1-Z_2)$

and, as $Q=V_2\times a_2$,

$$Q=a_2\left[1-\left(\dfrac{a_2}{a_1}\right)^2\right]^{-1/2}[2g(h_1-h_2)+(Z_1-Z_2)]^{1/2} \qquad [12.1]$$

If the two pipe areas are known and, if the pipe is installed horizontally, this expression becomes:

$$Q=\left(\begin{array}{c}\text{a numerical function}\\ \text{of the area}\end{array}\right)\times\left(\begin{array}{c}2\times g\times\text{pressure head}\\ \text{difference}\end{array}\right)^{1/2}$$

However, the assumption of negligible energy losses between the two sections is not quite accurate, and a coefficient – the **coefficient of discharge** (Cd) – has to be inserted. This coefficient is the ratio of

$$\dfrac{\text{the actual flow rate }(Q)}{\text{the flow rate predicted by [12.1]}}$$

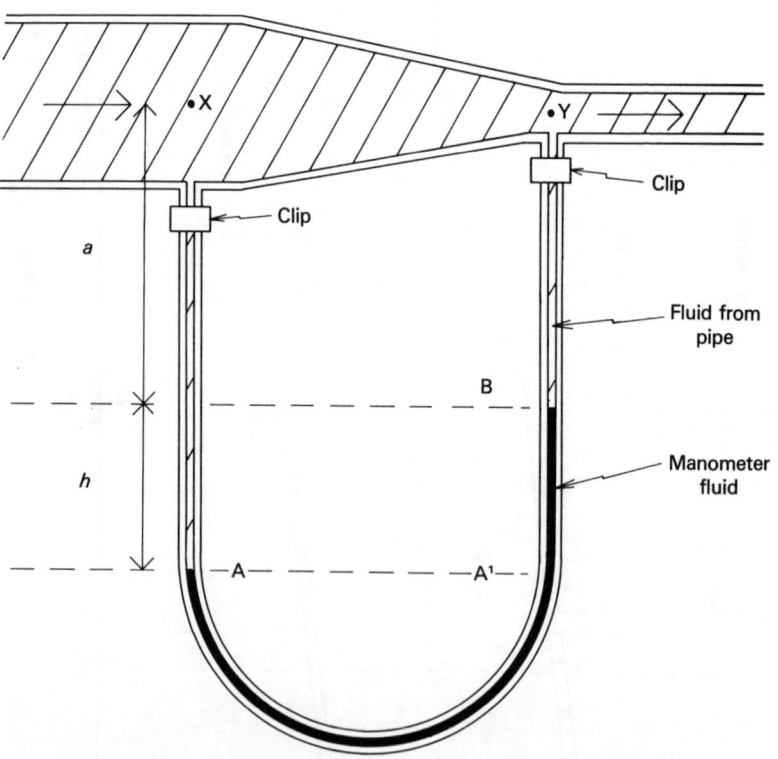

Fig. 12.2 Differential U-tube manometer

and has values from 0.97 to 0.61 for the commonest devices of this type. Some variation of Cd does occur with Reynolds numbers less than 10^5.

The general form for the flow equation of a horizontal pipe flow meter is thus

$$Q = Cd \times \begin{pmatrix} \text{numerical function} \\ \text{of areas} \end{pmatrix} \times \begin{pmatrix} 2 \times g \times \text{pressure head} \\ \text{difference} \end{pmatrix}^{1/2} \quad [12.2]$$

The coefficient of discharge may be determined by a simple laboratory calibration, where known flow rates are passed through the meter.

Thus the problem of measuring pressure pipe flows is reduced to being able to measure the pressure head difference, accurately, at two points. The U-tube manometers and pressure transducers of Ch. 2 are suitable for this purpose, though use is often made of a **differential mano-meter** (Fig. 12.2).

Such gauges' results are analysed exactly as were those of the simple U-tube manometers.

The pressure difference between points X and Y is required and it is known that the pressures at A and A^1 are the same. Thus

$P_A = P_X + \rho_1 \times g \times (a + h)$

and $P_A^1 = P_Y + \rho_1 \times g \times a + (\rho_2 \times g \times h)_{\text{gauge fluid}}$

Thus $P_X - P_Y = g \times h(\rho_2 - \rho_1)$

A variety of tapered section flow gauges exists, and the commonest are the **Venturi meter** and the **orifice plate**.

A Venturi meter (Fig. 12.3a) is a contraction of pipe section, followed downstream by a gradual return to the original pipe area. This downstream taper is included to minimise turbulent eddy production, and with good design, using a 6° angle of taper, only some 10 per cent of the velocity head is lost in eddies. The coefficient of discharge is thus high (usually about 0.97).

The necessary precision of manufacture and the required instrument length make Venturi meters expensive as pipe fittings. Thus their use is generally restricted to situations where energy losses must be minimised.

Where these circumstances do not prevail, a cheaper and simpler device – the orifice plate – is preferred. This is merely a circular disc, with an orifice machined in its centre. The disc is easily bolted between pipe lengths (Fig. 12.3b).

Flow through an orifice plate differs in two important respects from that in a Venturi meter. First, and most obvious, is the zone of turbulence downstream of the orifice. This causes a large energy degradation and leads to a coefficient of discharge of about 0.61. The other difference is that the smallest cross-sectional area of flows occurs, not at the orifice, but a short distance downstream of it, because of fluid inertia. On the face of it, this last point should create a real problem, since measuring the cross-sectional area of a vena contracta is far from simple,

232

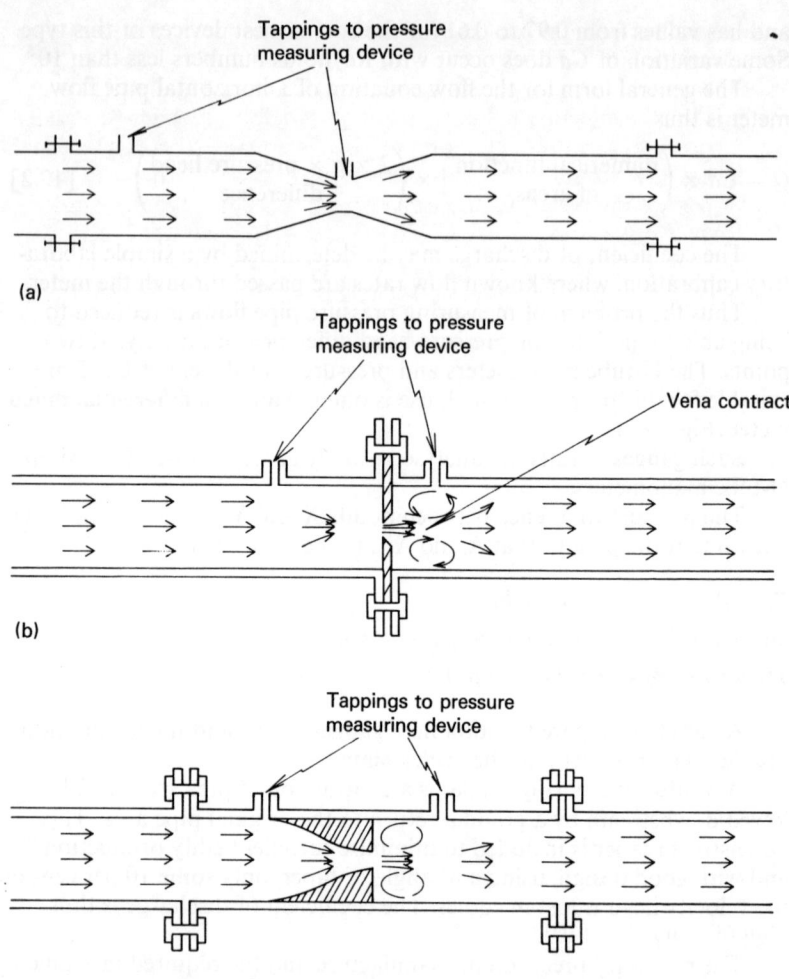

Fig. 12.3 Flow measuring devices for pressure pipe. (a) Venturi meter (b) Simple orifice plate (c) Streamlined orifice

and the term a_2 ([12.1]) does refer to this area. In fact, the problem is easily avoided by taking the area of the orifice as a_2 and inserting yet another coefficient – the **coefficient of contraction** (C_c) in the discharge equation – i.e. for a horizontal orifice plate,

$$Q = Cd \times C \times \begin{pmatrix} \text{numerical function} \\ \text{of areas} \end{pmatrix} \times \begin{pmatrix} 2 \times g \times \text{pressure head} \\ \text{difference} \end{pmatrix}^{1/2}$$

[12.3]

Devices with properties between the two commoner types have also been developed (Fig. 12.3c).

Flow meters of the types shown in Fig. 12.3 can be made to fit any pipe and to measure any flow rate to a high order of accuracy (normally to ± 1 per cent). The only limitation on their use arises from the fact that the highest flow velocity, and thus the lowest pressure head, occurs in the narrow part of the meter. If this area is made too small, then cavitation (Ch. 10) will ensue, and the gauge will cease to give accurate results.

Apart from these hydraulic devices, various mechanical meters are widely used. Usually, they have propellers, vanes, discs or pistons which are moved by the fluid and so operate a dial or a digital counter. The normal domestic or commercial water meter is of this type. Although they are cheap and convenient they suffer, as do all mechanical gauges, from wear and reduced accuracy with time. Thus they are not to be recommended as prime measuring systems, and should only be installed where an approximate flow rate is all that is required.

Sections 12.1 to 12.2 – Self assessment questions

1. From any suitable library text give examples of cases where routine flow measurements are needed.
2. Explain the theoretical basis for measuring the flow rate in pressure pipes.
   Detail precisely the meaning of the coefficients C_d and C_c in such methods.
3. Decide where a Venturi meter and an orifice plate should be installed.
4. Why is a differential manometer installation to be preferred to that of two standard U-tube manometers?

12.3 Measurement of flow in open channels

In open channel flow, no measuring system is available which approaches the cheapness, convenience and accuracy of the tapered section pipe gauges. This, of course, stems largely from the greater variation in velocity that occurs in open channels (Fig. 11.4).

Instead, three quite distinct approaches are employed:

1. Weirs.
2. Velocity-area gauging.
3. Chemical dilution.

12.3.1 Weirs

A weir is an obstruction, installed in a channel, to back up the fluid and force it to flow over, or through, the obstruction. The resultant critical depth of flow (Ch. 11) then allows a simple derivation of the required flow rate.

The theory developed in Chapter 11 (Fig. 11.6) shows that only the critical depth of flow is uniquely related to the discharge over a weir, and

234

it certainly would be expected that this is the depth which would have to be measured. In practice, however, measuring critical depth is not a straightforward task, since the instability of the critical flow allows very minor variations in the incoming specific energy (S.E.) to lead to quite large depth changes.

To avoid this, the flow depth is measured a short distance upstream of the weir, before the downdraw curve has developed (point (1), Fig. 12.4).

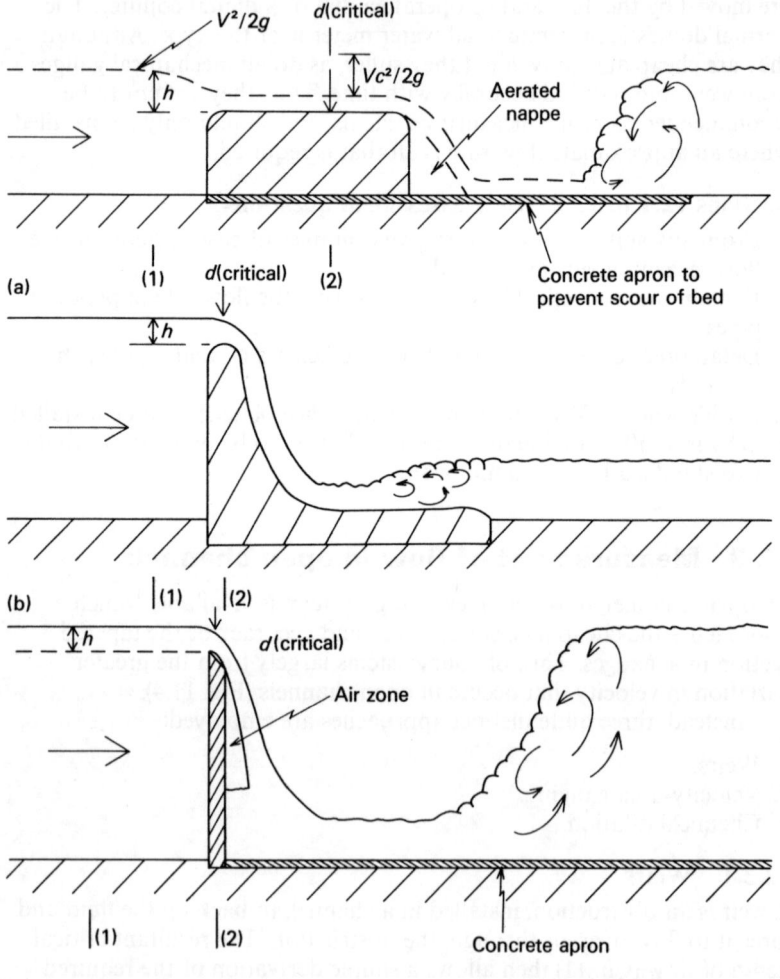

Fig. 12.4 Types of weirs. (a) Broad-crested weir (b) Round-crested weir (c) Thin plate weir

If the distance upstream is short and if, in line with the treatment of Chapter 11, a rectangular channel is assumed, then if

the specific energy at point (1) = that at point (2)

S.E.(1) = the critical specific energy at point (2)

$$= \frac{3}{2} \times \text{critical depth of flow (from [11.7])}$$

$$\text{S.E.}(1) = \frac{3}{2}\left(\frac{q^2}{g}\right)^{1/3} \qquad\qquad [11.5]$$

As the upstream depth of water (h) is large compared to that of the critical depth, the upstream velocity is likely to be small, and the velocity head there ($V^2/2g$) is probably insignificant. The total critical specific energy, therefore, over the weir, is approximately equal to the upstream water depth (h).

The flow rate, per unit channel width, can thus be expressed as:

$$q = \left(\frac{2}{3}\right)^{3/2} (g)^{1/2} (h)^{3/2}$$

or the equation for the total flow over the weir is of the form,

$$Q = C \times L \times h^{3/2} \qquad\qquad [12.4]$$

which is the general weir equation, applicable to all types of weir.

The term L is the crest length of the weir and the coefficient C is included to correct the error caused by assuming that the upstream velocity was insignificant.

As with pipe flow meters, the coefficient has to be determined by laboratory calibration, and it should be noted that it is not a constant but varies, to a small extent, with the upstream head.

The method of determining channel flow with a weir, is to measure this upstream head, usually by means of an autographic water level recorder, installed in a float well, to prevent the float being affected by the channel's current (Fig. 12.5).

The recorded heads can then be translated into discharge units quite simply, though computer assistance is usually necessary because of the large amount of data to be analysed. Such autographic recorders, and their more modern punch tape successors, can measure the water depth as frequently as once per minute, throughout a period of many hours or indeed days.

A variety of weir types are available and include:

thin-plate weirs – rectangular and triangular notches;
broad crested weirs;
round crested weirs;
flumes;
crump weirs;
flat-vee weirs.

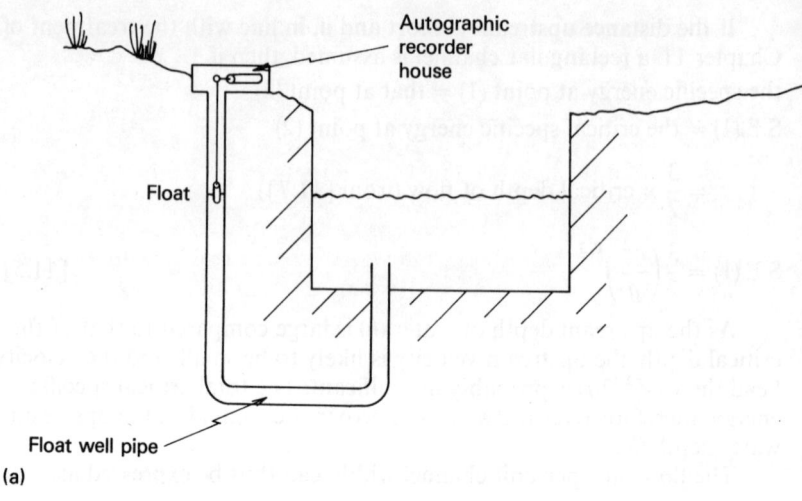

Float

Float well pipe

(a)

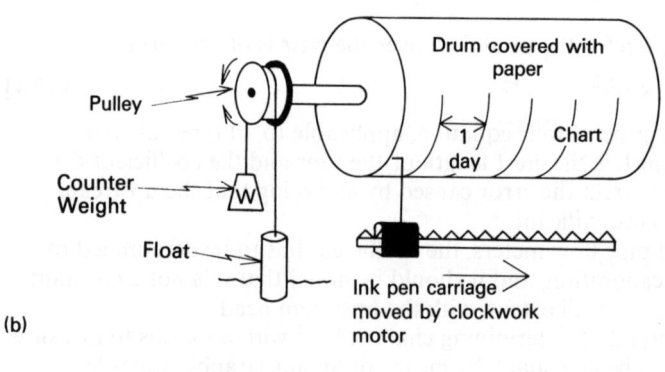

Pulley

Counter Weight

Float

(b)

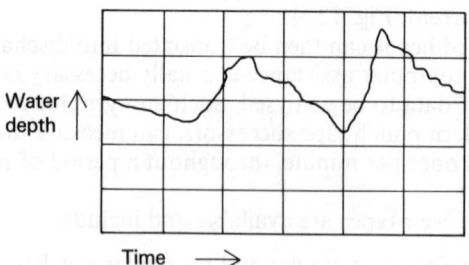

Water depth

Time

(c)

Fig. 12.5 Measurement of head over a weir. (a) Float well for measuring water depth upstream of weir (b) Simple autographic recorder (c) Autographic chart showing water level changes

Thin plate weirs (Fig. 12.4 c) are little more than steel plates, inserted vertically into a channel section and with a crest plate of some non-rusting metal (brass is usual) with a precisely machined bevel. The opening in such weirs can be made triangular, for accurate measurement of small flow rates, or rectangular, when larger flow rates have to be passed.

The accuracy achievable with thin plates is by far the highest of any type of weir (about 1–2 per cent in a precisely rectangular channel), but their use is restricted to laboratory channels, simply because they are unable, structurally, to withstand the forces met in real life situations and their crest units – the accuracy of machining of which is crucial to accurate operation – are easily damaged by floating debris and vandalism.

Broad-crested weirs have already been outlined in Chapter 11. Essentially these are low dams across the entire channel, which – to produce critical depth over the weir – must have a width, in the direction of channel flow, of at least three times the maximum height of water (h – Fig. 12.4a) above the weir. Thus they are massive and expensive structures, which invariably collect silt against their upstream faces and require annual maintenance, if accurate flow readings are required. As measuring systems they also tend to be inaccurate at low rates of flow, when the frictional resistance of the necessarily long width of crest slows the oncoming flow considerably.

The round-crested weir (Fig. 12.4b) obviously uses less concrete in its construction, and should be a cheaper proposition. However, the crest and the downstream slope have to be shaped precisely to the form of a free-falling sheet of water and thus a high, and expensive, quality of finish is needed. As with broad crested weirs, these tend to silt up and need maintenance.

Flumes (Fig. 12.6) also produce critical depth, by reducing the cross-sectional area open to the water flow, but do so by altering the channel width and not the bed level.

They do not suffer from siltation problems, since any sediment is washed through the structure. However, their length, and thus costs of construction, are higher than for normal weirs, set at right angles to the channel flow. To minimise these costs, a large number of satisfactory flumes have been constructed by driving sheet piles into the channel bed and backfilling to the original river bank with any suitable fill materials.

The equation for flow through a flume can be derived by a Bernoulli analysis of the conditions at its inlet and throat. This gives

$$Q = \frac{C \times W_1 d_1 \times w_2 d_2 \times (2g)^{1/2}}{W_1^2 d_1^2 - w_2^2 d_2^2}(d_1 - d_2)^{1/2}$$

which can be simplified to

$$Q = 1.71 \times (C)(C_v) \times W_2 \times H^{3/2} \qquad [12.5]$$

The additional coefficient (C_v) is included to allow for a slight variation, from the theoretical distribution of velocity, in the flume inlet.

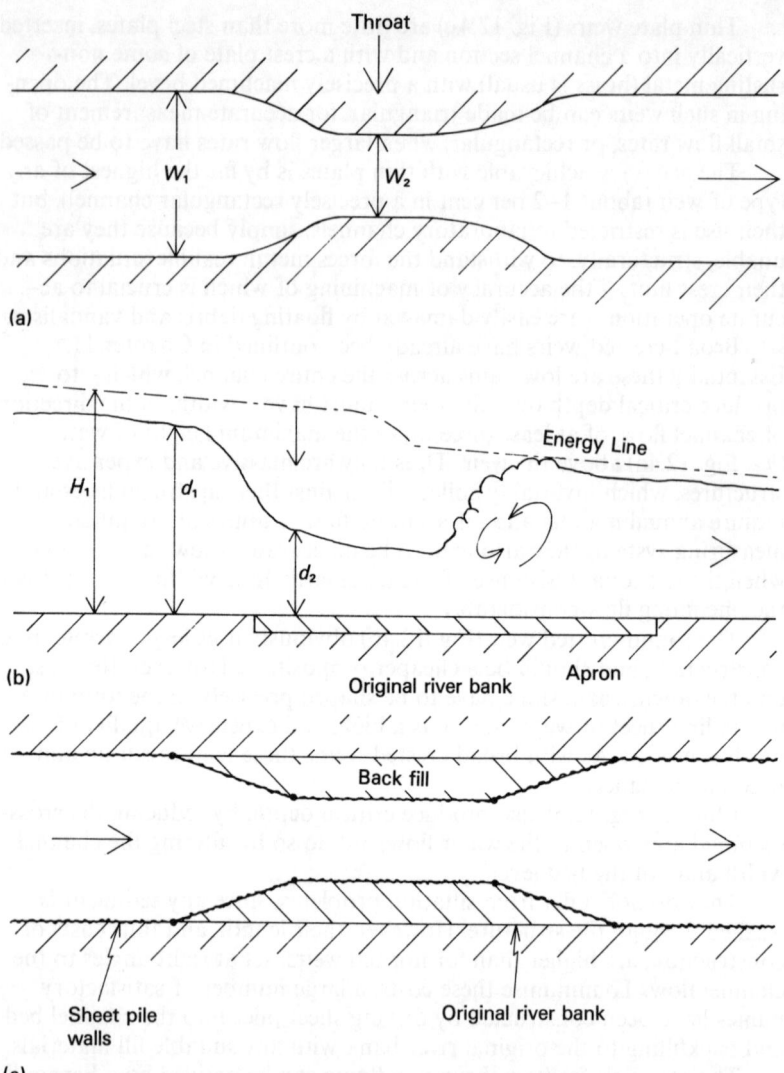

Fig. 12.6 Flumes for open channel flow measurement. (a) Plan view (b) Section (c) Plan of sheet pile flume

Apart from being free from siltation problems, the great advantage of a flume is the very limited backing up of the upstream water level it produces. In flat areas, this is a considerable advantage and has ensured that flumes are still amongst the commonest of measuring weirs.

Crump and flat-vee weirs are modern developments, designed to increase flow measurement accuracy and to reduce constructional costs.

The original crump weir is triangular in section (1 to 2 upstream and 1 to 5 downstream slopes) and requires less constructional materials, as well as giving a wider range of flows which can be accurately measured. The flat-vee is a crump cross-section, with a V-shaped flow area. To a very large extent, this combines the non-silting characteristics of a flume with the accuracy and lower cost of a crump weir.

As all weirs are intended to create critical flow in normal operation, a hydraulic jump usually occurs on the downstream side. The turbulent motions of the water in a jump make an excellent erosion tool and if a protective concrete apron is not provided (Figs. 12.4a, 12.4c and 12.6b) then the weir can be undermined and made structurally unstable. Even for quite small weirs a protective apron is required and often a raised sill is installed on its downstream side to provide a water-filled **stilling basin** to dissipate the effects of the hydraulic jump.

Flow equations – some of which have been listed above – have been derived for every established type of weir, and it might be thought that, if a weir is built in a channel, it will always provide the correct flow rates (subject to its own accuracy limitations). In fact, this is often not the case, and a need for periodic calibration (usually by velocity-area methods) is quite common. The reason for this is that the calibration of many weir designs took place in regular sectioned laboratory channels, and conditions in real open channels, where a more variable geometry and greater velocity variation occur, can differ markedly from the calibration environment. Additionally the siltation that can affect many weir designs can, in channels carrying large loads of suspended material, reduce the accuracy of the measured flows, to the point where the measurement is almost unjustified.

The choice, calibration and control of flow measuring weirs is thus a specialism in its own right and expert guidance should always be sought if a non-specialised organisation decides to enter open channel flow measurement using weirs.

12.3.2 Velocity area gauging

With weirs, the geometrical and velocity variations in open channels are minimised, as far as possible, by building expensive structures, the costs of which usually amount to several thousands of pounds even for quite narrow channels.

A cheaper solution, and one which does not attempt to modify the flow conditions in the channel, is the **velocity-area** method of gauging.

The theoretical basis of this method is (Fig. 12.7) to split the channel's cross-section into blocks of fairly uniform bed and roughness conditions and to measure the average velocity of flow in each of the blocks. Since the total flow (Q) equals the sum of the flows in the chosen blocks,

$$Q = q_1 + q_2 + q_3 + + + + q_n$$

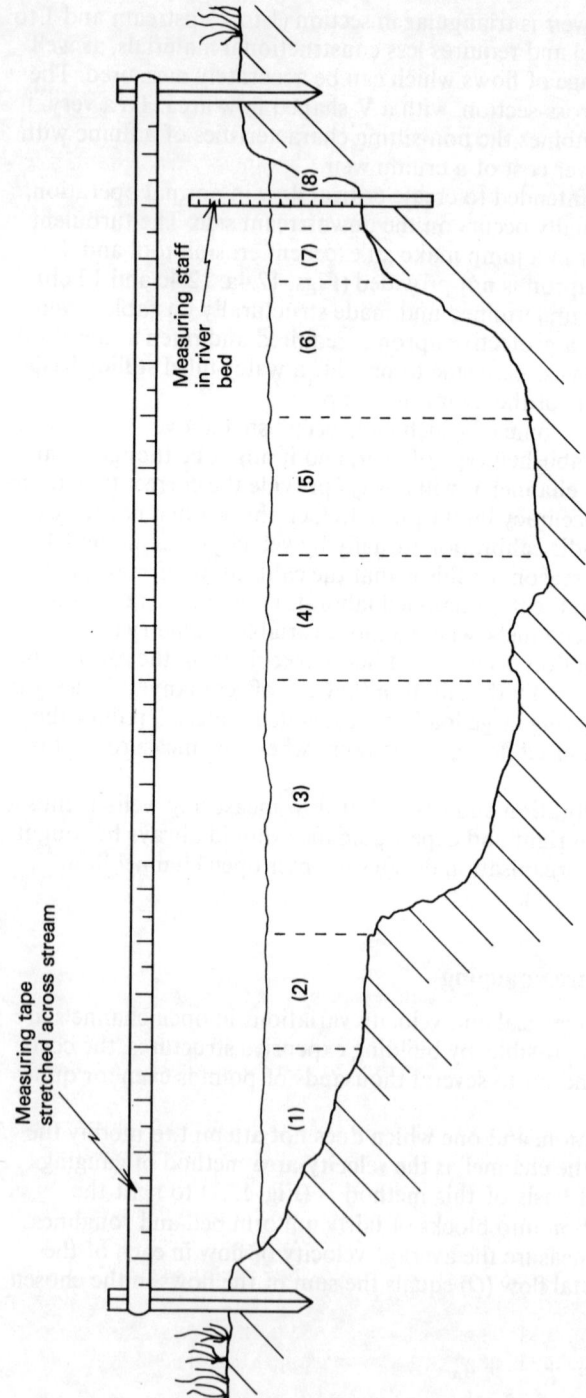

Measuring tape stretched across stream

Measuring staff in river bed

(1) (2) (3) (4) (5) (6) (7) (8)

Fig. 12.7 Velocity–area gauging of small stream

(where q_n is the flow rate in block number n)

$$= \bar{V}_1 a_1 + \bar{V}_2 a_2 + \bar{V}_3 a_3 + + + \bar{V}_n a_n \qquad [12.6]$$

The eight blocks, chosen in Fig. 12.7, have approximately uniform bed conditions, and so should not have too wide a velocity variation. It is noticeable that they are not all of the same width, as it is more important to have uniform velocity than constant width.

The gauging procedure, for a small stream, is therefore:

(a) install a measuring staff in the stream bed;
(b) stretch a survey tape tightly across the stream;
(c) measure the channel depth at intervals across the stream and produce a cross-sectional profile of the bed;
(d) choose the blocks (here (1) to (8)) of relatively constant velocity;
(e) knowing the width (from (b) and (d)) of the first block and its average depth of flow (from (c)), a current meter is then used to measure the average velocity. This occurs in the middle of the block, at 60 per cent of the total depth (measured from the water surface) or is more exactly given by the average of the velocities at the 20 per cent and 80 per cent depths (also measured from the surface);
(f) when the velocity in block (1) is determined, enough data is available to resolve the expression

$$q_1 = a_1 \times \bar{V}_1$$

and the current meter can be moved into the middle line of block (2), and so on;
(g) As the average velocity in each block is determined, the water level on the measuring staff should be noted to ensure that the flow rate in the channel is not varying, as the gauging is proceeding. Flows in natural channels do vary with rainfall and the rate of their variance can be surprisingly rapid. If the water level does alter during a gauging, there are methods of still achieving a measured flow of some accuracy, but, in general, this is not recommended, and the gauging should be abandoned;
(h) the total discharge is then calculated from [12.6].

Whilst the procedure is quite straightforward, one point does require further elaboration. The measurement of the stream's depth, at different points across its section, is actually quite difficult to achieve. The measuring rod will sink into soft mud, unless an enlarged base plate is installed, and where loose boulders floor a channel it is almost impossible to obtain an accurate measurement of water depth.

The current meter, whose employment is essential in velocity-area gauging, is a small (100–150 mm long), streamlined torpedo-shaped instrument with tail propellors, which are turned by the water's velocity. This rotation makes and breaks an electrical contact in the meter and, in turn, this operates a digital counter, which the operator carries above the water. The current meter is usually screwed on to the measuring rod, used earlier to obtain the water depths, and the meter is simply moved to the 60 per cent depth position and exposed to the water current for periods of about one minute.

With care, a gauging of this type will yield a flow rate accurate to about ± 5 per cent.

Velocity-area gauging methods can also be used for continuous measurement of the flow in a channel. In such cases, an autographic recorder and float well are installed and periodic velocity-area gauging is carried out to relate the various possible flow depths to discharge values. Usually in such cases, the channel will be too deep for a wading gauging of the type described above and some variety of cableway to move the current meter across and through the channel's cross-section will probably be required.

In all cases where velocity-area gauging is proposed, it is essential to have as long and straight a channel section as is possible to produce the minimum velocity variance.

The one situation where this method is not suitable is when the water in a channel is highly turbulent, since then the blades of the current meter may be damaged, and certainly will receive unbalanced forces, when struck by transverse water currents.

In summary, velocity-area gauging, to be successful, needs the same conditions of uniform river flow and lack of turbulence that weirs also need. The major difference between the two methods is one of capital cost, since the accuracy of results in each case is about the same.

12.3.3 Chemical dilution

The condition of highly disturbed rough water, which defeats both weirs and velocity-area gaugings does occur, and sometimes in places where a measurement of the flow rate is required.

In such circumstances, the only suitable technique is that of **chemical dilution**.

The basis of the method is that if a flow rate of Q litres per second exists and if the water has a concentration (of any particular dissolved chemical) of C grammes per litre of the water, then the weight of that chemical in the stream is

$$C \times Q$$

Thus if a small, known volume of a high chemical concentration is trickled into the stream, and mixed with it, by the water turbulence, a sample of the stream water taken further downstream, when analysed, will show an increased concentration of the chemical (Fig. 12.8).

Taking the river flow as Q, its natural concentration of the chosen chemical as C_0, the injection rate of the solution at (1) as q, its chemical concentration as C_1, and the concentration found after analysing water samples from (2) as C_2, the required river flow rate can be calculated

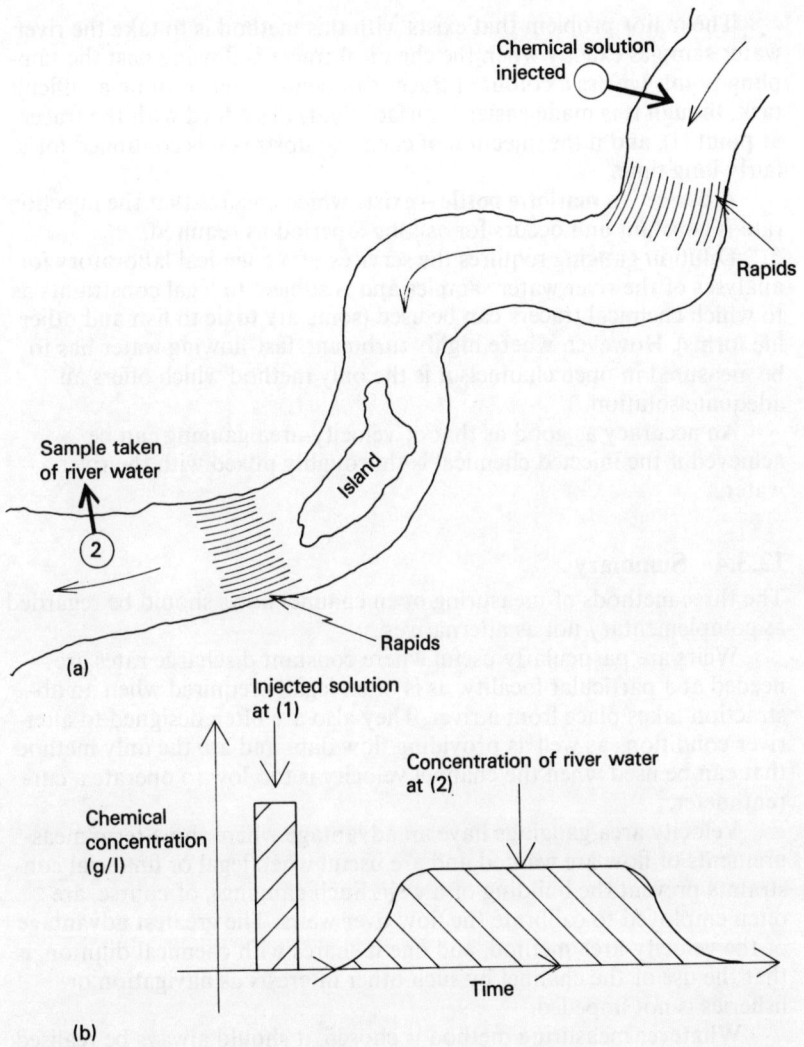

Fig. 12.8 Chemical dilution gauging. (a) Plan (b) Variation in water chemistry

thus:

(since the total weight of the
chosen chemical at point (1) = that at point (2))

then $\quad Q \times C_0 + q \times C_1 = (Q + q) \times C_2$

or $\quad Q = q \dfrac{(C_2 - C_1)}{(C_0 - C_2)}$ [12.7]

The major problem that exists with this method is to take the river water samples exactly when the chemical tracer is flowing past the sampling point. Unless a coloured tracer can be used, this is quite a difficult task, though it is made easier if surface floats are added with the tracer, at point (1), and if the injection of chemical upstream is continued for a fairly long time.

A device – a **mariotte bottle** – exists which ensures that the injection rate is constant and occurs for as long a period as required.

Dilution gauging requires the services of a chemical laboratory for analyses of the river water samples and is subject to legal constraints as to which chemical tracers can be used (some are toxic to fish and other life forms). However, where highly turbulent, fast-flowing water has to be measured in open channels it is the only method which offers an adequate solution.

An accuracy as good as that of velocity-area gauging can be achieved if the injected chemical is thoroughly mixed with the river water.

12.3.4 Summary

The three methods of measuring open channel flows should be regarded as complementary not as alternatives.

Weirs are particularly useful where constant discharge rates are needed at a particular locality, as is often legally required when an abstraction takes place from a river. They also are often designed to alter river conditions as well as providing flow data and are the only method that can be used when the channel velocity is too low to operate a current meter.

Velocity area gaugings have an advantage where short-term measurements of flow are wanted and are useful when legal or financial constraints prevent the building of a weir. Such gaugings, of course, are often employed to calibrate the flow over weirs. The greatest advantage of the velocity-area method, and one it shares with chemical dilution, is that the use of the channel by such other interests as navigation or fisheries is not impeded.

Whatever measuring method is chosen, it should always be realised that the difficulty of accurate measurement of open channel flow is such that specialist staff will be required.

Section 12.3 – Self assessment questions

1. Explain why the depth of flow, which should theoretically be measured over the crest of a weir, is normally measured a short distance upstream. How is the inaccuracy, of assuming that the velocity head (upstream of the weir) is negligible, overcome in the general weir formula?

2. Detail the usual layout of float well and automatic water level recorder needed if a continuous measurement of a river flow is to be obtained.

3. From a suitable library text, list the various types of weir available

and assess their relative merit in terms of cost, accuracy of measurement and suitability for various open channel situations.

4. Explain why a downstream apron (and often also a stilling basin) should be installed with a flow measurement weir.

5. What is the theoretical basis of the velocity-area gauging method? List the procedures necessary if such a measurement is to be taken on a stream, shallow enough for wading.

6. From a suitable library text obtain exact details of a current meter, and explain how the device allows a measurement of the average velocity of flow in a channel.

7. What are the conditions needed for an accurate velocity-area gauging?

8. Detail the theoretical basis of the chemical dilution method.

9. Consider all three open channel flow measurement methods and list their respective advantages and disadvantages.

Examples

Example 12.1

A horizontal Venturi meter has inlet and throat diameters of 300 mm and 100 mm respectively, and a coefficient of 0.98.

When a particular water flow is passing through the meter, a 65 mm difference is noted in the mercury levels of the differential manometer attached to the Venturi meter's pressure tappings.

What is the flow rate?

Solution

The areas of the inlet and the throat are

$$a_1 = \pi \times \frac{0.3^2}{4} = 0.0707 \text{ m}^2$$

and $$a_2 = \pi \times \frac{0.1^2}{4} = 0.00785 \text{ m}^2$$

Thus the numerical function of the areas in [12.2] is **0.0079 m²**. The pressure head loss across the meter is

$$P_x - P_y = g \times h (\rho_{mercury} - \rho_{water})$$

and, in terms of metres head of water, the pressure head difference is

$$0.065 \times 12.6 = 0.819 \text{ m}$$

Thus,

$$2 \times g \times \text{head difference} = 16.069$$

and the flow rate of water through the meter is

$$Q = 0.98 \times 0.0079 \times (16.069)^{1/2}$$
$$= 0.031 \text{ m}^3/\text{s}$$

Example 12.2

If the Venturi meter (in Example 12.1) is 500 mm long and if it is installed vertically (with the water flow direction down the meter), what is the flow of water?

(Assume the same dimensions and differential gauge reading as in the previous example.)

Solution

The only difference from the earlier example is that the term $(Z_1 - Z_2)^{1/2}$ in [12.1] can no longer be ignored.

As the flow is vertically down the pipe, the required flow rate is

$$Q = 0.98 \times 0.0079 \times [16.069 + (Z_1 - Z_2)]^{1/2}$$
$$= 0.98 \times 0.0079 \times [16.569]^{1/2}$$
$$= 0.0315 \text{ m}^3/\text{s}$$

Example 12.3

A 100 mm diameter orifice is installed in a horizontal 200 mm diameter pipe, in which a flow (0.020 m³/s) of water takes place.

If the overall coefficient of the meter ($C = Cd \times Cc$) is 0.600, what difference in mercury levels in a differential manometer will occur?

Solution

From [12.3], the required flow rate is

$$Q = Cd \times Cc \times (\text{function of areas}) \times (2 \times g \times h)^{1/2}$$

The areas are

$$a_1 = \pi \times \frac{0.2^2}{4} = 0.0314 \text{ m}^2$$

and $\quad a_2 = \pi \times \dfrac{0.1^2}{4} = 0.00785 \text{ m}^2$

Thus the function of the areas is **0.00809** and the term

$h = \textbf{0.865} \text{ m}$

= the difference in mercury levels in the two limbs of the manometer.

Example 12.4

The discharge from a laboratory channel is measured by a thin plate weir with a rectangular flow passage 0.300 m wide.

(a) how would the flow formula for this weir be found?
(b) what discharge would occur if a depth of 200 mm of water is above the weir crest and the discharge coefficient is 0.600?

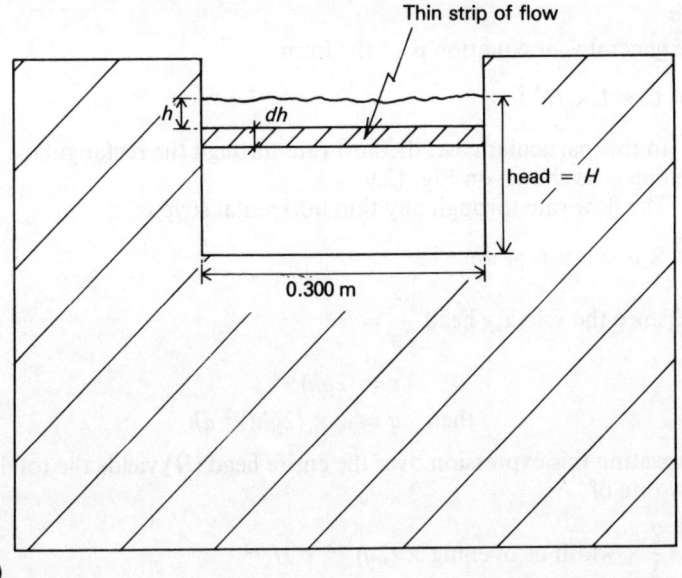

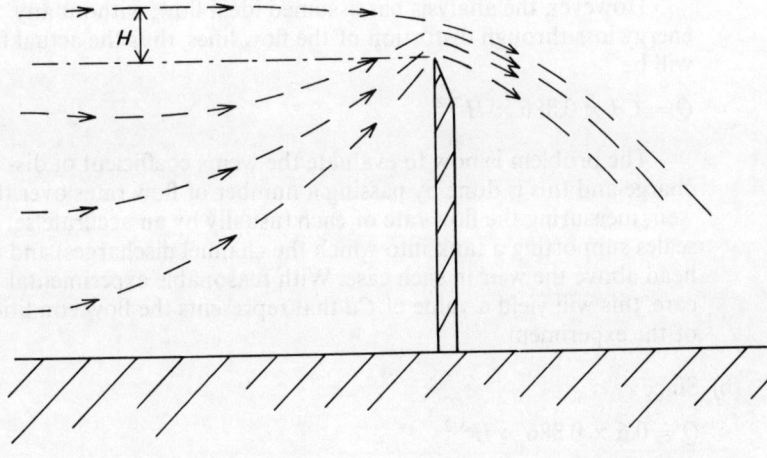

Fig. 12.9 Flow through a rectangular thin plate weir (Example 12.4). (a) View of weir from downstream position (b) Side view of weir

Solution

(a) The general weir equation is of the form

$$Q = C \times L \times H^{3/2}$$

In this particular case, the flow rate through the rectangular opening is as shown on Fig. 12.9.

The flow rate through any thin horizontal strip is

$$= v \times a = v \times w \times dh$$

and since the velocity head $\dfrac{v^2}{2g} = h$

$$v = (2gh)^{1/2}$$
$$\text{then} \quad q = w \times (2gh)^{1/2} \, dh$$

integrating this expression over the entire head (H) yields the total flow rate of

$$Q = \frac{2}{3} \times \text{width of opening} \times (2g)^{1/2} \times H^{3/2}$$

and, in this particular case of an opening 0.300 m wide,

$$= 0.886 \times H^{3/2}$$

However, the analysis has assumed ideal flow, without any energy loss through distortion of the flow lines; thus the actual flow will be

$$Q = Cd \times 0.886 \times H^{3/2}$$

The problem is now to evaluate the weir's coefficient of discharge and this is done by passing a number of flow rates over the weir, measuring the flow rate of each (usually by an accurate set of scales supporting a tank into which the channel discharges) and the head above the weir in each case. With reasonable experimental care, this will yield a value of Cd that represents the flow conditions of the experiment.

(b) Since

$$Q = 0.6 \times 0.886 \times H^{3/2}$$

if the head is 0.200 m then the flow rate = **0.0475 m³/s**

A flow measurement station is required on a river 300 m downstream of a works, where material from coal tips is washed to recover the coal from waste material. The effluent from the works is returned to the river.

The river flows through a low lying landscape and so has a smooth flow. Which weir type would you propose and for what reasons? Would any other measuring system be preferable to a weir?

Solution

The effluent from a washery plant will carry large concentrations of suspended solids. These, if allowed into slow moving water will settle (see Fig. 11.17) and create silt banks. Any type of weir which is a dam across the entire stream will have such slow-moving water on the up-stream side. Siltation upstream of a weir will cause inaccurate flow measurements, as the water depth, at any flow, will be raised higher as the silt bank grows. Thus the depth–discharge relationship of the weir will vary with time.

If a weir has to be built in this situation, then it should be of a self-cleansing type.

The only two options of this variety available are a flume or a flat-vee weir.

Velocity-area gauging is unlikely to be useful in this situation, since the growth of silt banks in any areas of reduced river velocity would cause the depth conditions for a particular flow to be variable.

Chemical dilution would be unsuitable as the river is stated to be non-turbulent.

The choice of a weir – despite the problems likely to be associated with it – is thus the best option in this case.

Example 12.6

A velocity-area gauging produced the results shown in Table 12.1.
What is the discharge in the river?

Table 12.1

Distance from R.H. bank (m)	Width of section (m)	Average depth of sectors (m)	Measured velocity at 60 per cent depth (m/s)
Block 1			
0–1.5	1.5	0.3	0.62
Block 2			
1.5–1.82	0.32	0.42	0.71
Block 3			
1.82–2.73	0.91	0.51	0.82
Block 4			
2.73–3.16	0.43	0.58	0.93
Block 5			
3.16–4.71	1.55	0.62	1.27
Block 6			
4.71–4.92	0.21	0.4	0.65
Block 7			
4.92–5.10	0.18	0.35	0.51

Solution

Assuming that the blocks can be treated as rectangular in shape, the discharge through each is

$$= \text{width} \times \text{average depth} \times \text{average velocity}$$

Thus   (1) = 0.279 m³/s
     (2) = 0.095 m³/s
     (3) = 0.381 m³/s
     (4) = 0.232 m³/s
     (5) = 1.220 m³/s
     (6) = 0.055 m³/s
     (7) = 0.032 m³/s

Thus the total discharge = **2.294 m³/s**

Answers

Chapter 1

4. Hydrostatics; hydrodynamics
Hydrostatics considers only a single type of force; hydrodynamics is complicated by additional forces

6. Not for hydrostatics, where the theory is adequate, but it certainly is true for hydrodynamics, where the ability to visualise what will happen is crucial for any understanding

Chapter 2

Section 2.1

1. Pressure intensity
2. 596.475 N
3. $P = \rho \times g \times h$
   $= 9810 \, \text{N/m}^2$, or
   $= 9.81 \, \text{kN/m}^2$
4. (a) $9.810 \times 10^{-3} \, \text{N/mm}^2$
   (b) 98.1 millibars (since 1 millibar $= 10^2 \, \text{N/m}^2$)
5. Yes
   (a) $P = 176.58 \, \text{kN/m}^2$
   (b) $P = 180.00 \, \text{kN/m}^2$
6. 34.66 m
7. Total pressure = total weight
   $$= 50 \, \text{kg} \times \text{gravity}$$
   $$= 50 \times 9.81$$
   $$= 490.5 \, \text{newtons}$$

$$\frac{\text{pressure}}{\text{intensity}} = \frac{490.5}{\text{area of plate}}$$
   $$= 49.05 \, \text{kN/m}^2$$
   $$= \text{pressure intensity transmitted}$$

8. (a) 15.941 kN/m^2
  (b) water 915.028 kN
    oil   686.274 kN
  (c) water 163.827 kN/m^2
    oil   147.886 kN/m^2
9. 1665.31 m

Chapter 3

1. No
2. Yes, in the special case of a plane subjected to uniform pressure intensity
3. Lower
4. Yes, since $F = A \cdot \rho \cdot g \cdot \bar{h}$
6. No, this is a unique relationship for rectangular bodies
7. Yes
8. Yes, but if an axis of symmetry exists it is always easier to project areas on to this axis

Chapter 4

1. Displace its volume of water from a tank into a measuring cylinder
3. Buoyance force is simply the reduction, in air weight, a body experiences when submerged, wholly or partly, in a fluid. As Example 4.2 shows, this reduction exactly equals the difference in pressure forces on the body's lower and upper surfaces
4. Because pipes, from time to time, have to be emptied
6. With the development of off-shore oil and gas exploration, another civil engineering speciality has arisen which requires a greater application of buoyancy effects

Chapter 5

1. No, on both counts
2. Kinematic viscosity has units of m^2/s
3. Velocity variation across stream, frictional loss against solid surfaces, type of flow that occurs, minor (eddy) losses and drag forces experienced by solid bodies in the fluid
4. Thicker
5. No
6. A property of both
7. Yes. Growth of weeds in a river in summer will give a rougher bed and so more friction losses. The dying off of these weeds in winter will lead to a smoother bed and so smaller frictional losses
8. Turbulent flows can look highly disturbed with rough surfaces and obvious disturbance; however, in many other cases the flow looks

quite smooth. The basic point is that turbulent flow is where the head losses are related to a higher power of the average velocity. The common English meaning of the word 'turbulent' is misleading here as it does suggest that eddies have to be present

9. No, unless the depth of water was very great
10. Anywhere where the flow paths are disturbed. Around bends and through valves are common examples

Chapter 6

Section 6.1

1. The term 'steady flow' means that the flow rate in a length of pipe or channel does not change with time. Inflow is thus equal to outflow. 'Unsteady flow' is the opposite of this
2. The flow in a river constantly changes in response to rainfall and typifies unsteady flow
3. No, only to steady flow

Section 6.2

3. To the hydraulic grade line which marks out the pressure head through the pipe
4. No. Where the pipe contains a sub-atmospheric fluid pressure, the line lies below the pipe
5. Always at a distance of $V^2/2g$ m above the hydraulic grade line
6. Yes. The loss of energy due to viscous drag on the walls of the pipe or channel always ensures that the total energy falls off with length of flow

Section 6.3

2. Those situations where fluid energy losses, between the points of interest, cannot be quantified are appropriate for the momentum equation

Chapter 7

Section 7.2

1. It is used to accelerate the fluid into the pipe and to create the full bore velocity of flow
4. No. Section 7.3 shows that the head loss in fluid flow varies with the discharge being passed
5. Pressure head
6. No. The use of $K \times V^2/2g$ is only a convenience to express a particular number of metres of head lost. Analytical methods (see section 7.3) invariably allow you to calculate the velocity head and so express minor losses

Chapter 8

2. The average length of the initial (k_o) and ultimate (k_t) roughness elements on the pipe's inner walls; millimetres; H.R.S. design chart, Chapter 6
5. 1009.7 m
8. Flows into junctions positive, flows out of junction negative
9. Pipe from A = 1.183 m³/s
   Pipe to B = 0.321 m³/s
   Pipe to C = 0.862 m³/s
11. The higher the C value, the lower the flow that can be developed

Chapter 10

Sections 10.2 to 10.4

2. By ensuring that the cross-sectional areas of the plunger and the ram are in the correct ratio
3. Yes
4. By a development of surge pressures on valves of carefully chosen weight
5. Yes
6. Yes. One method commonly used was to install several pistons and suction chambers in one pump body
7. As shown on Fig. 10.4
8. All these devices depend on valves and pistons which have to be water and air tight. Earlier materials (usually leather) aged and needed replacement. Today a variety of more durable and smoother materials exists

Section 10.5

1. The desired form of hydraulic energy is pressure head, which is needed to power pipe flow. The high velocity head produced by a centrifugal pump has to be converted to pressure head and the expanding flow area in a scroll case does this
2. See section 8.5.1
3. Yes, since fluid power produced = $\rho \times g \times Q \times H \cdot$ efficiency
4. Yes
5. Yes, by placing the pump far enough below the source water level. The advantage is that cavitation effects can be ignored
6. The head, discharge and power values at the maximum efficiency point
7. Because most fluids contain dissolved gases and these separate (to give cavitation problems) at suctions rather smaller than the vapour pressure of the fluid
9. Altitude (i.e. atmospheric pressure), temperature of the fluid being pumped, and the suction head losses
10. Yes. These factors reduce the possibility of cavitation effects

13. Because the cheapest possible pipe is invariably accompanied by a very expensive pump, and vice versa. Thus the best combination of pipe diameter and pump has to be calculated

Sections 10.6 to 10.7

2. Because at low discharge rates, the power demands of an axial flow pump are at their greatest, and if the flow rate is cut back (by a control valve) the risk of overloading the pump motor exists
4. The high cost of fuel oils has made the high capital costs of hydro-electric schemes much more acceptable

Chapter 11

Sections 11.1 to 11.3

1. With open channels, the H.G.L. coincides with the water surface, the wetted perimeter (and so friction) vary with the quantity of the discharge, the pressure head is simply the water depth and the velocity distribution is less precisely known than it is in pipe flow
2. Yes
3. The depths of the water at various points in the channel. Without this, it would not be possible to ensure that flooding is avoided
4. See Fig. 11.2; in the middle reaches of straight lengths of channel of constant cross-sectional area
5. No. Calculate the value of normal depth from the Manning equation (assuming you can estimate n and also measure the discharge rate accurately) and check whether this depth corresponds to that in the middle of the length of channel
6. In open channels, the velocity distribution is less precisely known and this precludes the development of more precise design tools. Additionally there are greater economic pressures to size pipes exactly than is the case with open channels

Section 11.3

1. The energy of an open channel flow relative to the channel bed. It differs from the normal Bernoulli energy value only in the omission of the elevation term (Z)
2. Figure 11.6 makes it clear that quite small variations in the specific energy (from the critical depth value) can lead to quite large changes in the water depth in the channel. Thus a channel designed to flow at critical depth could experience uncontrolled depth variation if the specific energy were altered (by, for example, the annual growth of water weeds)
3. No, only to rectangular sections
5. Because the additional specific energy required for a smooth transition is unavailable and yet the bed slope cannot support the super-critical flow. The condition is thus unstable and leads to a standing wave disturbance in the channel

6. A downdraw curve implies a reduction in water depth down the channel; a backwater curve is the reverse situation
7. Yes
8. The backwater profile is, of course, a curve, and [11.10] is that of the straight line. To produce the necessary precision, it is essential to split the backwater curve into small portions, over which the straight lines, defined by [11.10], accurately predict the water profile

Chapter 12

Sections 12.1 and 12.2

1. The examples could include: the monitoring by the relevant authority of the abstractions taken by a farmer from a river; the measurement of discharges being put into public sewers by commercial companies, who have to pay for this service; the leak detection surveys commonly carried out on a town's water mains in the night-time period when consumer demand is miniscule
3. Venturi meters reduce turbulent eddies and so energy degradation. Their use is preferred where fluid energy is thus scarce or where it has been bought to power a pump. Orifice plates are cheaper to buy and install but degrade more energy. Their use is best where energy losses are of little consequence
4. The differential device is cheaper and more compact, and simpler to read than are two separate instruments

Section 12.3

1. By insertion of the coefficient C, whose magnitude is obtained from a laboratory calculation
7. Long straight channel reach without visible turbulence

Index